Geography, Geology, and Genius

Geography, Geology, and Genius: How Coal and Canals Ignited the American Industrial Revolution

by

Martha Capwell Fox

Geography, Geology, and Genius:
How Coal and Canals Ignited the American Industrial Revolution

Copyright © 2019 Delaware & Lehigh National Heritage Corridor

All rights reserved. No parts of this book may be reproduced or used in any form or in any manner whatsoever without the express written permission of the publisher, except for the use of brief quotations in critical articles and reviews.

Photographs and drawings used in this book are copyright © the persons and institutions that gave permission for use of images from their collections.

Published by Canal History and Technology Press
an imprint of the Delaware & Lehigh National Heritage Corridor

https://delawareandlehigh.org/

2750 Hugh Moore Park Road, Easton, PA 18042

Contact archives@delawareandlehigh.org for permissions

ISBN: 978-0-930973-44-5

Printed in Pennsylvania

Cover image:

"Down Among the Coal Mines—Weighing the Cargoes in the Weigh-Lock on the Lehigh Canal"
Wood engraving by Paul Frenzeny published in *Harper's Weekly*, February 1873
(*NCM/D&L collections, photographic copy by Paul C. Fox*)

DELAWARE & LEHIGH
NATIONAL HERITAGE CORRIDOR
CANAL HISTORY & TECHNOLOGY PRESS

This publication has been made possible through a generous grant from
Furthermore: a program of the J.M. Kaplan Fund

Furthermore:
a program of the J.M. Kaplan Fund

Dedication

This book is dedicated with great affection and respect to Lance E. Metz,
retired historian for the National Canal Museum,

and

to the generations of workers
in the industries that made the Corridor into what it is today.

Contents

Acknowledgments v
Foreword vii
Introduction xi

1. 1790–1819: Coal Ignites the Fire 1
2. 1820–1830: Coal and Canals 13
3. 1830–1840: Connecting Mine to Market 23
4. 1840–1860: Revolutionary Anthracite Iron 35
5. 1840–1850: The Iron Age 45
6. 1850–1860: Mineral Wealth 57
7. 1860–1870: Iron Horses and Iron Rails 73
8. 1870–1880: Rise and Fall and Rise 85
9. 1880–1890: Steel, Cement—and Silk 105
10. 1890–1900: Armor, Immigration, and Cement . . 117
11. 1900–1910: New Century, New Industries, and a New Name 133
12. 1910–1920: Labor War and World War 155
13. 1920–1930: The Roaring Twenties 169
14. 1930–1960: Depression, War, Booms, and Busts . 187

Epilogue 199
Author's Note 205
Endnotes 207
Bibliography 215
Index 219

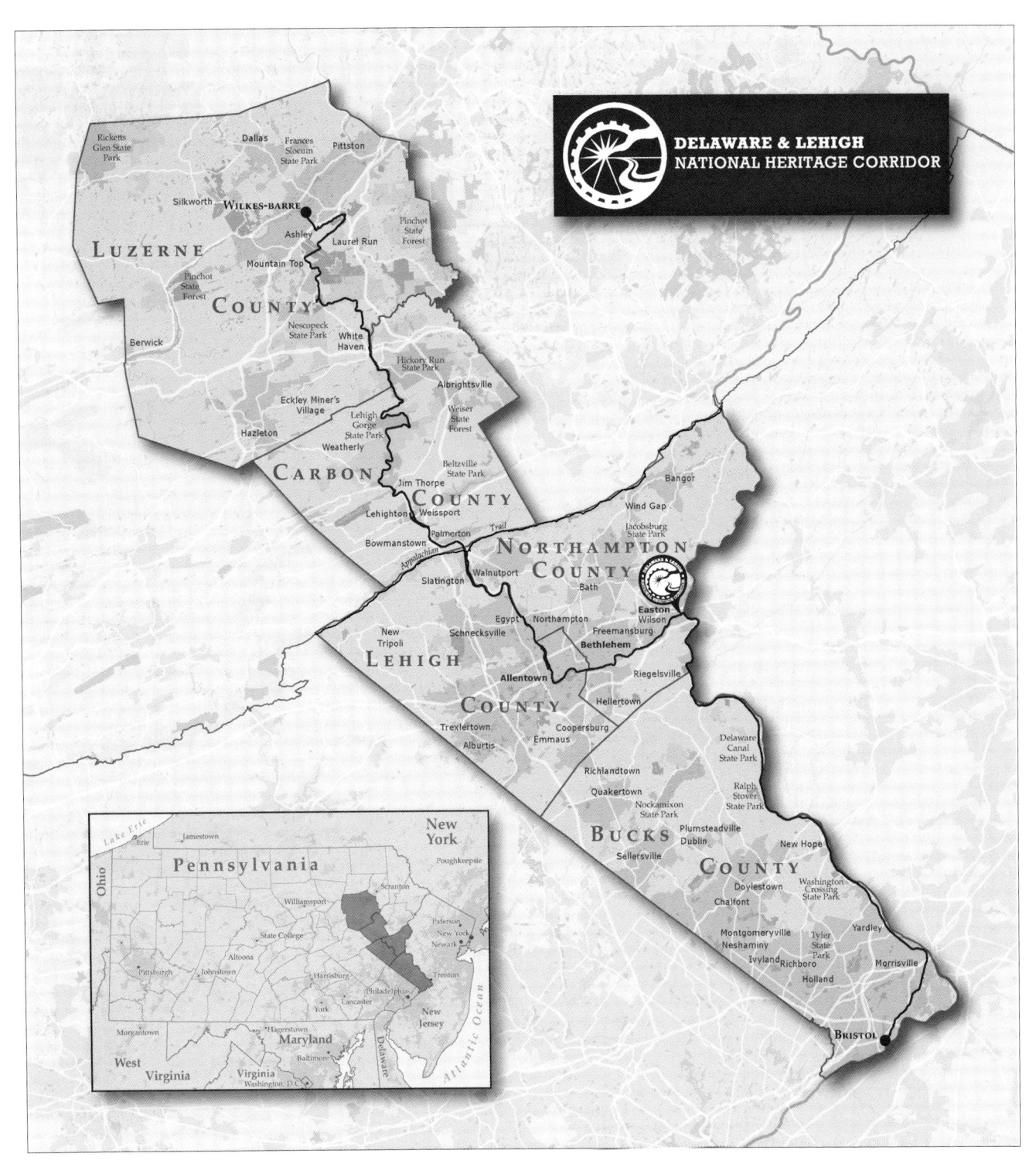

The counties of the Delaware & Lehigh National Heritage Corridor, showing their common element—the Lehigh and Delaware rivers—and the Corridor's location in the Commonwealth of Pennsylvania.

Foreword

In 1988, the United States Congress established the Delaware & Lehigh National Heritage Corridor, forever changing what would become the comprising five counties. Since then, the D&L has honored its commitment to preserving and interpreting the nationally significant history of the Corridor.

A region of rich natural resources and industrial distinction, the Corridor has long been a nucleus of innovation and growth. Mining and entrepreneurship shaped the Corridor throughout the eighteenth and nineteenth centuries; iron and anthracite coal, two of the largest industries at the time, created abundant jobs and fostered prosperity. With the help of the auspiciously located Lehigh and Delaware rivers, the region utilized canals and mule-drawn boats to ship hundreds of tons of goods to nearby Philadelphia and New York, propelling the Industrial Revolution and contributing to the development of our country.

Today, the Corridor looks much different. But only by understanding the past can we understand the present and look toward the future. This is the philosophy of the Delaware & Lehigh National Heritage Corridor. The rich history of the five counties, with their industrial, cultural, and natural interest, offers a wealth of material to help us understand our place in an ever-changing world.

Anticipating a dynamic future and keeping with the growth-centered legacy of the Corridor, the D&L plays a fundamental role in ensuring that the economic engine of the region continues to churn. Each year, it generates $474.7 million in economic impact and supports 5,665 jobs. We are continuing the transformation of the former route of coal and merchandise into what will be the longest multi-use trail in Pennsylvania, stretching from Wilkes-Barre to Bristol. Along it, you can connect with the history shared in this book and follow in the footsteps of both mules and mankind—and add your own chapter to a legacy that assuredly will extend far into the future.

Michael Drabenstott

Chair, Board of Directors
Delaware & Lehigh National Heritage Corridor
January 2019

Acknowledgments

A BOOK ABOUT HISTORY is never written by only one person. This is very much the case with this history of the industrial development of the D&L National Heritage Corridor.

In 2013, Tom Stoneback, then executive director of the National Canal Museum, and I rediscovered the 369-page *Historic Resources Study* that was prepared in the late 1980s to support the proposal to the National Park Service and the U.S. Congress for creation of the Delaware & Lehigh National Heritage Corridor. This was a comprehensive documentation of the national significance of the history of five counties that make up the Corridor and the sites, records, and institutions that preserved and presented that history. The *Study* began with the early contacts in the Corridor between the Native Americans and European settlers, covered the events of the American Revolution that occurred in the Corridor, and, most importantly, traced the development of the American Industrial Revolution in the nineteenth century that resulted from the discovery and efficient delivery of anthracite coal.

The proposal for the creation of the Corridor and the *Study* had both been spearheaded by the board of directors and the staff of Hugh Moore Historical Park and Museums, better known as the National Canal Museum. The effort involved many local and regional governments and groups, while the *Study* itself was researched and written by several academic and architectural historians, some passionate though amateur local historians (including this writer), the Bucks County Conservancy, and the Wyoming Historical and Geological Society. It is a sweeping, monumental, detailed telling of the story of the D&L Corridor, and it achieved its goal. An Act of Congress on November 18, 1988, established the Corridor, the second such designation of an entire region in the United States.

Then no one else read it. Though sweeping and detailed, the purpose of the *Study* demanded that it examine each area and industry individually, which prevented a synthesis of those histories into a coherent, readable story. Visitors to the Museum and the mule-drawn canalboat ride in Hugh Moore Park hear a digest version of the story, the D&L's *Tales of the Towpath* education program and Immersion Days field trips teach thousands of area fourth graders an age-appropriate version of it, and members of the D&L staff give talks on it to myriad adult groups. Oddly enough, in the twenty-first century, the D&L's nationally significant history has largely been told by the most ancient method—oral storytelling.

To rectify this situation, Tom Stoneback had the *Study* digitally scanned and asked me to assemble what initially looked like a fast cut-and-paste compilation of its industrial history parts.

Though the *Study* is widely quoted in this book, I quickly discovered that it alone would not tell the story we wanted. For one thing, in the late 1980s Bethlehem Steel was still (relatively) flourishing, so that one company accounted for nearly fifty pages. The *Study* was more than a little Lehigh Valley-centric—perhaps understandable at the time, but now not acceptable given the Corridor's well-established reach. Apart from a few maps and charts, the *Study* had no illustrations. It downplayed or overlooked entirely some very interesting developments and communities. Worst of all, from my perspective, there was little attention paid to *people*.

What is history other than the stories of people who, though they came before us, actually were just like us? Like us, they lived, worked, slept, ate, learned, suffered, worried, dreamed, and loved. Some of them, the ones we remember because their names are on buildings, schools, and streets, achieved great things. Because this is a book largely about heavy industry, those people are men who attained wealth and prestige through their efforts. But tens of thousands of others are equally important to our story, though their lives may be harder to draw out of the past, and their names are known only to their families or are forgotten. We can see them in early photographs and sometimes need only to look into their faces to hear their stories.

Historical images of buildings, industries, cities and towns also help tell the story, showing us what the Corridor our forebears lived and worked in looked like. Many of the images of people and places are from the collections held in the archives of the National Canal Museum and the D&L Corridor, including the Pennsylvania Canal Society collection. These are identified by the notation NCM/D&L. Others are used as noted, with the permission and courtesy of the following:

- Special Collections and College Archives, Diane Shaw, director, and Pam Murray, rare book cataloger, Skillman Library, Lafayette College
- The Moravian Archives, Bethlehem, Thomas McCullough, assistant curator
- National Museum of Industrial History, Bethlehem
- Bradley Pulverizer, Inc., David Fronheiser
- The Luzerne County Historical Society, Wilkes-Barre
- The Margaret S. Grundy Memorial Library, Bristol, Pennsylvania
- Mack Trucks Historical Museum, Doug Maney, Curator
- The Presbyterian Church of Catasauqua
- MMI Preparatory School, Freeland, Pennsylvania
- Lower Macungie Township Historical Society
- Atlas Cement Company Memorial Museum, Edward Pany, curator
- The Slate Belt Heritage Center
- Ron Bray
- Kelly Ann Butterbaugh
- Aaron Hackler
- Vincent Hydro
- Helen Dery Woodson and Cameron Smith

I'm grateful for the support and encouragement I've received for this project from my co-workers on the staff of the D&L, as well as for the financial support of the grant from the J.S. Kaplan Foundation.

Finally, the two people to whom I owe my deepest thanks are Ann Bartholomew and Lance Metz, for their dedication to our history, their willingness to share their incredible stores of knowledge, and their friendship. Ann was the editor of the *Historic Resources Study*, and she has brought her skill, diligence, enthusiasm, and creativity to dozens of history publications from Canal History and Technology Press and many other local history books. The book you are holding is every bit as much her work as mine.

Lance's wide-ranging intellect, prodigious memory, and energetic cultivation of a large circle of friends, colleagues, and contacts helped him broaden the original mission and collections of the National Canal Museum from a focus solely on canals to encompassing the entire story of anthracite and the American Industrial Revolution. From him, I—as well as many other people here and around the country—learned to appreciate the depth, breadth, and immense significance of the history of our region to the entire United States. It is this story that the Delaware & Lehigh National Heritage Corridor conserves and celebrates.

Martha Capwell Fox
Easton, Pennsylvania
March 2019

Introduction

Pennsylvania was the driving force of American industry in the nineteenth century. The Commonwealth's industries employed more people and produced more goods with more dollar value than any other state. Pennsylvania was the world's center of iron production, surpassing by the 1880s even the industrial might of Britain and Germany. And Pennsylvania was the nation's powerhouse, supplying 95 percent of the hot-burning, high-energy anthracite coal that ignited America's Industrial Revolution. That revolution began in the five counties—Bucks, Northampton, Lehigh, Carbon, and Luzerne—that are now designated the Delaware and Lehigh National Heritage Corridor. This is the place where America was built, the cradle of the American Industrial Revolution.

How did this slice of the Keystone State become the birthplace of modern America? The answer lies in the confluence of waterways, minerals, and minds that met here in the mountains and valleys bisected by the beautiful but narrow and rocky Lehigh River, and bounded on the east by the still-wild Delaware.

The key mineral was anthracite coal. Hidden beneath the mountains of northeastern Pennsylvania, a mammoth source of energy lay undiscovered for millions of years. Anthracite coal is purer, harder, and has a higher carbon content than any other coal. This means it burns hotter, cleaner, and longer than any other solid fuel. It is vastly superior to the fuels—wood, charcoal, dung, peat, straw, and even soft coal—that humans had used for millennia to cook their food, heat their dwellings, and make tools, implements, weapons, and ornaments from metal. When anthracite's energy was finally tapped here in eastern Pennsylvania, chemistry, metallurgy, industry, and agriculture—and all of American life—was changed forever.

So full of anthracite were these mountains that nearly all the stories of its discovery tell of someone simply finding it on the ground. The first people who were intrigued or persistent enough to get the coal to ignite found it made a hotter fire than any they had experienced. And since, as historian Henry Adams wrote, "The Pennsylvania mind … in practical matters it was the steadiest of all American types; perhaps the most efficient …" several people independently—in the Wyoming Valley, in the Schuylkill Valley, and in Philadelphia—decided to look into how this fuel could be used.

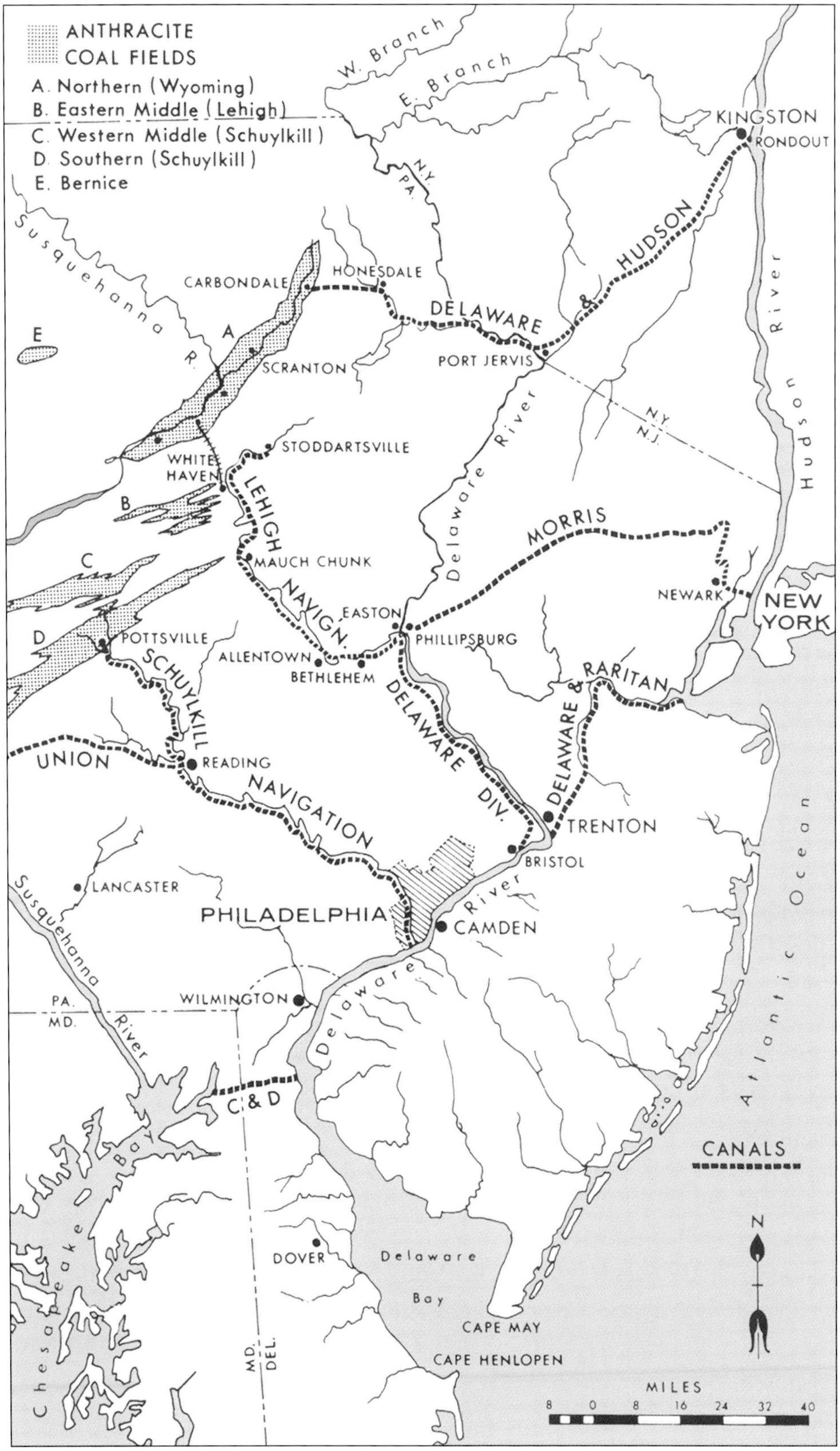
ANTHRACITE
COAL FIELDS
A. Northern (Wyoming)
B. Eastern Middle (Lehigh)
C. Western Middle (Schuylkill)
D. Southern (Schuylkill)
E. Bernice
W. Branch
E. Branch
KINGSTON
RONDOUT
N.Y.
PA.
HUDSON
&
DELAWARE
HONESDALE
CARBONDALE
Susquehanna R.
A
E
SCRANTON
PORT JERVIS
N.Y.
N.J.
Hudson River
Delaware River
STODDARTSVILLE
WHITE HAVEN
LEHIGH
B
MORRIS
MAUCH CHUNK
C
NAVIGN.
EASTON
NEWARK
NEW YORK
D
POTTSVILLE
PHILLIPSBURG
ALLENTOWN
BETHLEHEM
SCHUYLKILL
DELAWARE
DELAWARE & RARITAN
UNION
READING
NAVIGATION
DIV.
TRENTON
BRISTOL
LANCASTER
PHILADELPHIA
CAMDEN
Delaware River
Susquehanna River
WILMINGTON
PA.
MD.
Atlantic Ocean
C & D
CANALS
Chesapeake Bay
N
DOVER
Delaware Bay
CAPE MAY
CAPE HENLOPEN
MD.
DEL.
MILES
8 0 8 16 24 32 40

Using anthracite on a large scale meant getting it from the wild and remote mountains to Philadelphia. In the early nineteenth century, William Penn's "greene country town" was the financial, industrial, commercial, and intellectual center of the United States. A hundred miles to the north, however, the roads in those mountains were little more than Indian trails and since the Native Americans had mostly been driven out, almost no one lived there. Furthermore, the only way to move large quantities of any heavy, bulky cargo like coal was by water—and the only water available was two rushing, rocky rivers, the Schuylkill and the Lehigh, and the slightly less treacherous Delaware River, which the Lehigh empties into.

The earliest attempts to ship coal focused on building boats that were little more than rafts. These were seen as the only way to float the coal over the multiple obstructions the rivers, especially the Lehigh, threw in their way. While American minds everywhere from New England to New York to Virginia were turning toward harnessing the abundant power of the nation's rivers for their mills and factories, in Philadelphia two businessmen, Josiah White and Erskine Hazard, decided that a better idea would be to adapt the Lehigh River to their need to deliver coal to Philadelphia.

White, who was one of the most prolific American inventors of his time, focused on taming the rivers, while Hazard, the well-connected son of one of America's pioneers of the insurance business, courted investors and found the money. Once they had established a flourishing coal-mining and -delivery business, they set about bringing the nation into the industrial age, by fostering the change from using charcoal to using their anthracite to smelt iron. Between them, the two lifelong friends and business partners helped revolutionize transportation, industry, and business in the United States, and triggered industrial, technological and economic innovations that continue—in the Corridor, the Commonwealth, and the United States—in the twenty-first century.

These five counties are one of the places in the world where geology and geography met human genius, with momentous results. The biographer David McCullough writes "I [have never] been able to disassociate people or their stories from their settings, the 'background.' If character is destiny, so too, I believe, is terrain." The story of the Delaware and Lehigh National Heritage Corridor is how the terrain of eastern Pennsylvania, from the resource-rich wilderness mountains to the north to the educated cosmopolitan commercial environs of Philadelphia, coupled with the "Pennsylvania mind" and the extraordinary intelligence and imaginations, drive, and energy of thousands of native-born and immigrant Pennsylvanians triggered a social and economic transformation that changed the entire United States.

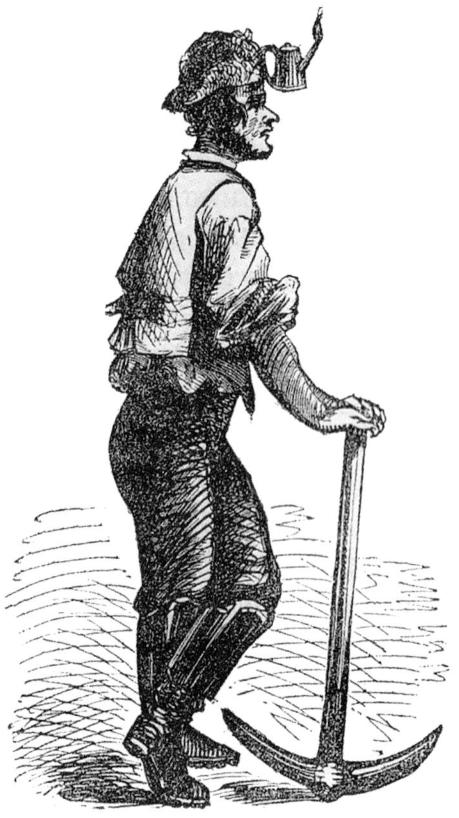

Chapter One

1790 TO 1819
COAL IGNITES THE FIRE

"I made diligent inquiry about the Coal on the Lehigh ..."

Josiah White's History given by himself, *1832*

During the twentieth century, most economic historians dated the start of the American Industrial Revolution to the 1810s, when water-powered textile mills were opened in New England. Then, in the early 1970s, Pulitzer Prize-winning business historian Alfred Chandler challenged that conventional wisdom. The birthplace of American industry was not New England, he maintained, but Pennsylvania—specifically, its anthracite regions and the valleys of the rivers that carried the coal to market.[1]

Chandler, born in Delaware but a New Englander for most of his life, demoted the early textile towns of his adopted home from their "birthplace" status by pointing out that production in the early mills was low, and usually seasonal. True, waterpower made America's first real factories possible, changing textile making from a home-based, part-time pursuit to an organized, centralized production process. But production was limited by factors like low water and winter ice. More importantly, textiles could not build a nation.

Nor would the "ancient techniques," as Chandler put it, that Americans of the eighteenth and early nineteenth centuries were using to make iron.

Most Americans of that time were well aware of the major technical advances in ironmaking—fueled by coal and coke—that had revolutionized industry in Britain during the previous seventy-five years. But American ironmakers had only wood, converted to charcoal, to fire their furnaces. Not only did this limit their production capacity to about twenty-five tons per week, at best, by the early nineteenth century, it was driving them far away from population centers where the forests had long since been cut down for farm fields, houses, shipbuilding, and firewood. By the 1820s, most of the iron produced in Pennsylvania came from three counties in the still rather remote middle of the state—Center, Huntingdon, and Bedford—where furnaces were located in the midst of forests that could be cut for charcoal.

Consequently, most East Coast manufacturers that needed iron—nail makers, coopers, blacksmiths, and gunsmiths—relied almost exclusively on iron imported from Britain. Even Pennsylvania's makers of metal-based products imported most of the iron they needed. This left them vulnerable to the price fluctuations and vagaries of trans-Atlantic commerce as Britain waged decades of war with France.[2]

Closer to the coast, where the nation's most populous cities were, and where most Americans lived, wood was becoming an increasingly costly commodity. Two centuries after the arrival of the first European settlers, most of the forests east of the Appalachians had been depleted by the need for lumber for building, heating and cooking, and making charcoal for ironmaking. Not just the ironmakers and their customers, but everyday Americans were being stressed by the increasing scarcity and rising costs of wood.

The other great innovation of Britain's Industrial Revolution in the eighteenth century that had failed to catch on in the United States was the coal-fired steam engine. The abundance of free water power in the ten northeastern and Middle Atlantic States where manufacturing was concentrated, and the absence of any combustible fuel other than increasingly expensive wood, made steam engines impractical except for powering boats. Besides, the primitive state of American metal-working technology made building large steam engines close to impossible.

What really gave life to the American Industrial Revolution? What powered Pennsylvania and the rest of the United States into the modern age? According to Chandler and other economic and industrial historians, the answer is anthracite coal.

"The importance of coal in American industrialization cannot be overstated," writes Walter Licht in *Industrializing America*.[3] Anthracite coal from Pennsylvania—one of the handful of places in the world where it can be found—almost completely displaced water, wood, and charcoal as America's energy sources in less than two decades in the middle of the nineteenth century. This set the nation on course for the explosion of economic, industrial, and social changes that altered the landscape not only of the Corridor and the Commonwealth at large, but of the entire United States.

The Discovery

The story of the discovery of anthracite coal is shrouded in legend, traditions, and debatable historical accounts. It is possible that Moravian missionaries reported the existence of anthracite coal deposits near the present Carbon County community of Summit Hill in 1746, and it is likely that anthracite from these deposits was burned in the forges of blacksmiths at the Northampton County Moravian settlement of Nazareth between 1750 and 1755. Tradition also asserts that the surveyor for the Susquehanna Company plotted an anthracite coal outcrop on a map recording his 1762 exploration of the Wyoming Valley.

In 1766 samples of Wyoming Valley anthracite were sent to Thomas Penn, proprietor of Pennsylvania, at London, and an early 1768 survey of what is now Plymouth Township in Luzerne County notes the existence of anthracite coal deposits in that area. The first fully documented use of anthracite coal in Pennsylvania occurred at Wilkes-Barre, in what is now Luzerne County, when two Connecticut blacksmiths, Obediah and Daniel Gore, used anthracite at their forge in

1768–1770. During the Revolutionary War small quantities of anthracite were shipped from the Wyoming Valley to the arsenal at Carlisle in 1776, and in 1779, officers of Sullivan's expeditionary force noted the existence of anthracite coal deposits. In 1788, Wilkes-Barre blacksmith Jesse Fell used anthracite at his primitive nail works.

These early attempts to discover and exploit the anthracite coal deposits within the region of the D&L Corridor predate the discovery and use of anthracite in other parts of Pennsylvania's anthracite coalfields. Not until 1784 was coal from the Schuylkill Valley mentioned in historical records, while the deposits of the Lackawanna Valley would not be exploited until after 1812.

Despite the possibility of the earlier discovery and use of anthracite from the Summit Hill deposits, the development of this coalfield dates from 1791, when a miller, Philip Ginder, discovered a commercially exploitable outcrop there. Stories vary about how and why Ginder actually made his discovery, but his subsequent action—taking a sample to Jacob Weiss—eventually developed the first commercially successful transportation link between the anthracite coalfields and a major metropolitan market, Philadelphia.

Artist's depiction of Philip Ginder's discovery of anthracite. (NCM/D&L)

Colonel Jacob Weiss was a local entrepreneur and landowner whose employees had been mining and using anthracite as blacksmiths as early as 1785. Weiss was impressed with the extent of the outcrop that Ginder had discovered and he believed that if anthracite from this location could be easily mined and transported to Philadelphia, it would find a ready market.

Weiss's assumption was based on the fact that, as a rapidly growing city of 42,000, Philadelphia had come to use bituminous coal as an alternate fuel to scarce and expensive wood and charcoal. By 1792 Philadelphia was importing over 1,000 tons of bituminous coal from Britain annually; smaller quantities of bituminous coal from Virginia's James River highlands also made their way to Philadelphia. Unlike the Wyoming Valley deposits, which were in the Susquehanna River watershed, the Summit Hill anthracite deposits were accessible from the Lehigh River, a tributary of the Delaware. Thus, coal from the location Ginder had discovered could reach Philadelphia via a water route, while coal from the Wyoming Valley could be shipped via the Susquehanna only to the relatively undeveloped areas of central Pennsylvania.

Col. Jacob Weiss

After securing the backing of capitalists from Philadelphia and the Moravian town of Nazareth, Weiss formed the Lehigh Coal Mine Company in 1792. Extracting the coal was not a prob-

Coal was originally quarried from surface deposits. After the outcrops of coal were exhausted, drift-mining techniques were used to extract coal from seams that went underground. This sketch of the Summit Hill pit is from Richard Richardson's 1873 Memoir of Josiah White. (NCM/D&L)

lem: the anthracite coal outcrops were close to the surface—as little as three to fifteen feet below the surface, so deep mining was unnecessary during the early years of the anthracite coal industry. At the Summit Hill open pit, quarrying techniques were used to extract the anthracite, and the mine yielded about 268 tons of anthracite every twelve-hour workday.

During the next decade this company successfully mined anthracite and transported it on arks via the connecting Lehigh and Delaware rivers to Philadelphia. However, the Lehigh Coal Mine Company proved to be an unprofitable enterprise for two reasons: the lack of a ready market, as most Philadelphians did not have stoves that were designed to burn anthracite, and the loss of many ark-loads of coal on the wild and hazard-filled waters of the Lehigh and Delaware rivers.

While the Lehigh Coal Mine Company was experiencing financial difficulties, attempts continued to exploit the Wyoming Valley coal deposits commercially. In 1807 two settlers from Connecticut, Abijah and John Smith, developed an anthracite mine in what is now Plymouth Township, Luzerne County. This mine became a major producer of anthracite coal, which was both marketed locally and shipped down the Susquehanna River. By 1822, the Smiths' company was shipping over 2,183 tons of coal annually, making it the pioneering firm in the commercial development of the Wyoming anthracite field. A large factor in the success of the Smith brothers' mine was Jesse Fell's development in 1808 of a home fireplace grate which made use of anthracite as a domestic fuel possible. Fell "advertised" his grate by burning anthracite at public demonstrations.

Earlier, in 1800 and 1804, similar devices had been devised by Philadelphia inventors Oliver Evans and Dr. Thomas James for their own use.

Energy Crisis Provides Opportunity

During the same period that the Smith brothers began the development of their Wyoming Valley mine, the Lehigh Coal Mine Company leased a portion of its coal-mine properties to the Philadelphia firm of Rowland and Rutland, but the lease was voided two years later when the firm failed at both mining and delivering coal. The outbreak of the War of 1812 brought the prospect of financial relief to the beleaguered Lehigh Coal Mine Company. The British blockade of the American coast had cut off the ship-borne supplies of British and Virginia bituminous coal on which Philadelphia's craftspeople, homeowners, and steamboat engineers had increasingly come to depend as fuel sources. By April of 1813 the price of a bushel of bituminous coal had more than tripled.

The scarcity and high price of bituminous coal made Philadelphians willing to experiment again with anthracite. The potential market in Philadelphia attracted the attention of several Wyoming Valley coal-mining entrepreneurs, among them Jacob Cist (1782–1825), the son of Charles Cist, a prominent Philadelphia publisher and member of the Moravian Church. He was also the nephew of the Lehigh Coal Mine Company's founder, Jacob Weiss. Charles Cist had been one of the original investors in the Lehigh Coal Mine Company and on his death in 1805, Jacob inherited his shares in this venture.[4] In 1807 Jacob Cist married the daughter of the Wyoming Valley's most socially prominent and wealthiest merchant, Matthias Hollenback. More than a son-in-law, Jacob Cist soon became his business partner as well, directly involved in his father-in-law's efforts to mine and market anthracite.

Self-portrait of Jacob Cist.

Cist and other Wyoming Valley entrepreneurs began hauling wagon loads of hard coal to Philadelphia in 1813 to supply fuel-starved Philadelphia craftsmen. Transportation by horse-drawn wagons over narrow, rough roads was then the only way to move coal from the mines to markets but, although the anthracite found a ready market, land transportation could not meet the demand in Philadelphia. The Wyoming Valley is over 120 miles north of Philadelphia, and it took almost a week for a team and wagon to transport less than two tons of anthracite to the Quaker City. Since the Susquehanna River flows south and west to central Pennsylvania and the Chesapeake Bay, Cist and the other Wyoming Valley coal entrepreneurs could not use the more efficient mode of water transportation to reach the Philadelphia market. Jacob Cist saw a new opportunity and turned his attention to developing the Lehigh Coal Mine Company's properties, from which coal could be shipped to Philadelphia via the Lehigh and Delaware rivers.

TO THE Blacksmiths OF THE CITY OF PHILADELPHIA.

The Subscriber having succeeded in overcoming the prejudice of his Workmen against the LEHIGH COAL, after two years solicitation and repeated trials, attended with disappointment, has had them used for 13 months past, to great advantage for himself and satisfactory to his journeymen (one of them that was the most adverse, is now using them at Boyertown, Berks County, at $75 per hundred bushels, in a neighborhood where Charcoal can be purchased for one-tenth of the sum)—now offers his advice and assistance *gratis*, to such Smiths as shall call on him, between the 13th and 19th of March inst. in altering their FIRE-PLACES to fit them for using said Coal. He will be found at the Bethlehem Stage Office, Sign of the Swan, Race Street, or the Seven Stars Tavern, New Fourth Street, above the Hay Market. The address of any person, desirous of obtaining information, left at either place, will be attended to. Gentlemen of the above business are informed that he is not concerned in the Coal speculation, but offers his assistance in aid of the profession, to fulfil a promise made the Coal Company last fall, which he could not heretofore attend to, owing to indisposition.

JOSEPH SMITH.

Philadelphia, March 13, 1815.

WE, the Subscribers, residents of the County of Bucks, Do Certify, that on the recommendation, and under the direction of Joseph Smith, we were induced to make trial of the Lehigh Coal, in our Smith-shops.—We have used them about four months; and believe, at the price we gave ($24 per ton) they are the most œconomical coals we could use. We find that the weight on the fire, the only objection to them, is more than compensated by the intensity of heat, and freedom from that corrosive quality and cinder, to which all other kinds of Coal are subject.

Given under our hands, February 24, 1815.

JACOB B. SMITH, *of New Hope.*
EDMUND KINSEY, *of Milton.*

The subscriber further says, that on a fair trial of the relative worth of the two kinds of Coal, he found that with 22 lb. of the Lehigh Coal, he could make 11 pair of horse shoes; and that it required 33 lb. of Richmond Coal to make the same number.—Time about equal; which, making the allowance for the specific gravity of the two kinds of Coal (the Richmond Coal being very dry) will make at least half the difference in measure; and that the Lehigh Coal fire was the most pleasant to work at.

March 1, 1815.

EDMUND KINSEY.

MINER'S PRESS—DOYLESTOWN.

On December 15, 1813, Jacob Cist and his partners, Wilkes-Barre entrepreneurs Isaac Chapman, Charles Miner, and John Robinson, obtained a ten-year lease of the Lehigh Coal Mine Company's anthracite lands in return for the production of 5,000 bushels of anthracite during each of the first two years of the lease and 10,000 bushels each subsequent year of the agreement. During 1814–1815 Cist and Miner attempted to create a market for Lehigh anthracite among the merchants and manufacturers of the Lehigh Valley and upper Bucks County. They achieved some success, and in particular sold significant amounts of hard coal to Quaker blacksmith Joseph Smith of Tinicum, Bucks County, a pioneer in manufacturing iron plows. Smith soon became an enthusiastic advocate of the use of anthracite as a forge and foundry fuel and issued handbills to Bucks County and Philadelphia blacksmiths endorsing its use, while at the same time he aided Cist and Miner by becoming a wholesale distributor of their product.

The development of a Lehigh Valley and Bucks County coal trade was only a sideline for Cist and Miner, who devoted much of their energies and capital to the shipment of anthracite from their mines to the Philadelphia market. During the spring of 1814, the partners managed to navigate several loaded coal boats, or arks, down the turbulent Lehigh and Delaware rivers to Philadelphia. However, many of their shipments were sunk by the rocks and rapids of the Lehigh River and the cargoes that did reach Philadelphia were sold at a price of approximately 50 percent of their fixed costs. By 1815, Cist and Miner realized that the arks they had been using were too large, with loads that were far too heavy to successfully navigate the Lehigh River, thus accounting for the loss of three out of every four coal cargoes that they shipped. To remedy this problem, they ordered the construction of smaller and more maneuverable boats, but their plans were frustrated by the end of the War of 1812 that winter. Once again, easily ignited bituminous coal could be shipped to Philadelphia from both Great Britain and the Chesapeake, effectively ending any market for Cist and Miner's anthracite and curtailing their enterprise.[5]

Enter White and Hazard

Josiah White and Erskine Hazard were Philadelphia entrepreneurs who had purchased anthracite from Cist and Miner in order to fuel their wire works at the Falls of the Schuylkill River near Philadelphia. They were among the most innovative of America's early transportation and industrial pioneers and their achievements would play a large role in the Lehigh Valley's contribution to the American industrial revolution.

The first major contribution of White and Hazard to the development of the use of anthracite coal was their accidental discovery of how to use it in a large-scale manufacturing capacity, which no one had yet done in the United States. A worker at White and Hazard's wire mill closed the door of a furnace that had been fueled with anthracite but had failed to heat up properly. When the worker returned a few hours later to retrieve a forgotten coat, the formerly recalcitrant furnace was glowing red hot. Josiah White was summoned and immediately reasoned that the closed furnace had created the conditions of a controlled draft, which allowed the anthracite to reach a high-enough temperature to combust.

This was anthracite's secret: it needs a controlled amount of air from underneath to ignite and burn. And when it burns, it gives off temperatures that had never been possible before. America, and the world, were on the brink of a revolution in metallurgy and chemistry.

This revelation awakened White and Hazard to the economic potential of anthracite coal as an industrial fuel. Since their factory was located on the Schuylkill River, which provided the shortest route from Philadelphia to the anthracite coal regions, White and Hazard turned their attention to the improvement of this natural waterway as a means of reaching the deposits of the upper Schuylkill Valley. They played a catalytic role in the organization of the Schuylkill Navigation Company in 1815, but they were soon forced out of this nascent enterprise by other capitalists who saw this proposed canal system as an integral part of a waterway to the developing west instead of merely a conduit for the shipment of anthracite to Philadelphia. Thwarted in their original intentions, White and Hazard turned their attention to the Lehigh coalfields as an alternate means of supplying the Philadelphia market.[6]

Owning the Lehigh River

In 1817, White and Hazard began their efforts to develop the coalfields of the upper Lehigh Valley. They sold their waterpower rights on the Schuylkill River to the City of Philadelphia, which then built the Fairmount Waterworks. Using this capital, White and Hazard began work in the Lehigh region. They leased the 10,000 acres of land owned by the Lehigh Coal Mine Company for an annual fee of one ear of corn, with the proviso that they were required to send, after three years, 40,000 bushels of coal annually to Philadelphia.[7]

Josiah White and Erskine Hazard

Josiah White (1781–1850) was born in Mount Holly, New Jersey, and apprenticed to the owner of a hardware business in Philadelphia at the age of 15. At 21, having learned the business, he bought his own store and set himself the goal of having $40,000 to his name by the age of 30. He achieved this in only five years, and sold the business to his younger brother Joseph and a partner. He then turned his attention to promoting manufacturing in the United States. In 1812, White used his wealth to open a wire and nail factory on the banks of the Schuylkill River with his business partner, Erskine Hazard.

Philadelphia-born Hazard (1790–1865), whose father was the first U.S. postmaster-general and a founder of North America's first insurance company, was raised in one of the best-connected families in the country, educated at Princeton (which had been founded by his grandfather), and published a scientific paper on electricity while still in his teens. Like White, Hazard invested his own wealth and used his family's connections to seek investors in the Lehigh Coal and Navigation Company. Hazard's keen, mechanically inclined intellect combined with White's inventive, business-oriented mind to overcome every challenge they faced in creating an industrial and transportation system unprecedented in the United States.

White and Hazard forged not only a strong, resilient business partnership that withstood cut-throat competition, a fraudulent associate, the rough-and-tumble of nineteenth-century Pennsylvania politics, and numerous natural disasters, but also a robust, respectful, and affectionate friendship that lasted until White's death in 1850. For the remainder of his life, Hazard faithfully monitored and guided the progress of the company they had built, trying to cope with the challenges of Asa Packer's railroad, the calamitous 1862 flood, and the specter of labor unrest. He was on his way to Mauch Chunk to deal with the murders of two company foremen, allegedly by the Mollie Maguires, when he died just days before the end of the Civil War.[8]

Josiah White's birthplace in Mount Holly, New Jersey.

(Richard Richardson, Memoir of Josiah White, *1873*)

Through the aid of a business associate, George Frederick Hauto (who later was revealed to be a scoundrel and a fraud), White and Hazard secured the passage of state legislation giving them the right to improve the navigation of the Lehigh River in 1818. In fact, this gave White and Hazard virtual ownership and monopoly control of the Lehigh River.

The legislators apparently had very little expectation that White and Hazard would fare any better than everyone else who had tried to make money from anthracite. As one said, the partners had been "given the privilege of ruining themselves." Dubious as they were that the enterprise would ever yield any money, the Commonwealth did not charge even a token ear of corn.

The sweeping provisions of the act gave them comprehensive rights to use and alter the river and its resources, including taking stone, gravel, and trees; to make any dam, lock, or other navigation devices; to enter and occupy all land necessary for contracting locks, sluices, canals, towpaths or other devices; to make bridges or fords; to collect tolls and duties on the navigation system so long as it was at least twenty feet wide and eighteen inches deep; and to use, sell, or rent the water or waterpower to any person, to be used in any way they thought proper. This made for a lot of bitterness and jealousy among White and Hazard's future business rivals. More immediately, it gave them a deadline: these rights would be forfeited if the lower navigation was not completed in seven years, and the upper in twenty years.

To secure additional capital, White and Hazard organized two separate companies in 1818: The Lehigh Coal Company, and the Lehigh Navigation Company. Their reasoning at the time was that few potential investors would be interested in funding both the coal-mining and navigation-improvement aspects of their operations.[9]

During 1818–1819 White and Hazard built a wagon road linking the coal deposits at the site of Summit Hill with the Lehigh River at the mouth of Mauch Chunk Creek. By bringing a crew of workers with them, living at first on boats moored near the worksites and eventually housing them, they founded the town of Mauch Chunk, now Jim Thorpe. The first house was built for Nicholas Brink, the steward of the Lehigh Coal and Navigation, and his wife Margaret, perhaps because they were expecting the first child to be born in the new settlement. Their son, born on April 21, 1819, was dubbed with the mighty moniker Josiah White Erskine Hazard George F.A.O. Hauto Brink.*

White solved the problem of navigating the Lehigh River by inventing hydrostatic or "bear trap" dams. These low dams had collapsible center sections that were normally held upright by the hydraulic pressure of the water. When the wickets (small iron doors controlled by upright rods that extended above the dam) alongside the center section were opened, the water pressure was released, the center section collapsed, and an artificial flood was created. This flood carried loaded coal arks from the pool behind the dam, over the now-flattened center section, and then downstream to the pool behind the next bear trap dam.[10]

---

* Josiah Brink, the name he went by, is buried in the GAR cemetery in Summit Hill; he died in 1878 at the age of 58. His tombstone records (though misspells) all of his names, so apparently neither he nor his parents ever relieved him of the burden of being named after the fake nobleman and swindler Hauto.

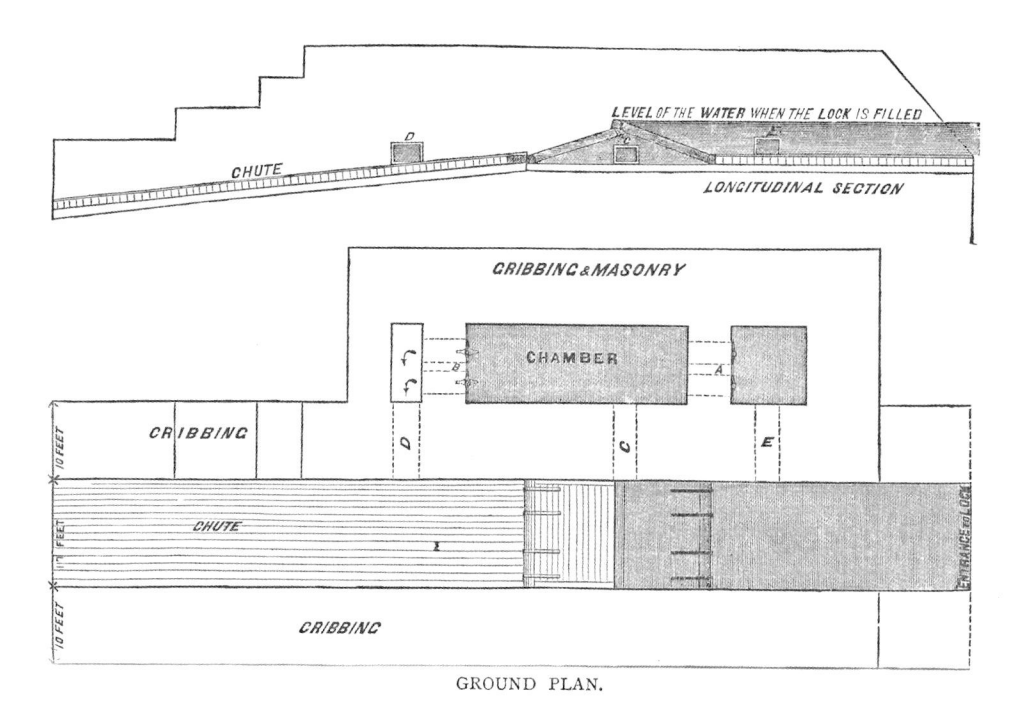

This sketch of a Bear Trap lock from Richardson's Memoir of Josiah White *is accompanied by a detailed description of how it worked (not included here).*

Despite problems caused by floods, droughts, the primitive wilderness conditions of the upper Lehigh Valley, and the exposure of their business associate, Hauto, as a bankrupt fraud, White and Hazard completed their descending navigation system on the Lehigh River in 1820. Within a year they had established a profitable and growing coal trade with Philadelphia.

The descending navigation system on the Lehigh River was the first commercially successful transportation link between the anthracite coalfields and Philadelphia. It immediately made large quantities of anthracite available cheaply, which in turn stimulated increased consumption and spread the use of anthracite among Philadelphia's residents and industries. Lehigh anthracite was exported from Philadelphia by coastal shipping companies to New York and New England, a trade that further stimulated the demand for anthracite.[11]

With the coalfields open and an efficient—though one-way—transportation system in place, White, Hazard, and their anthracite were poised to launch a revolution.

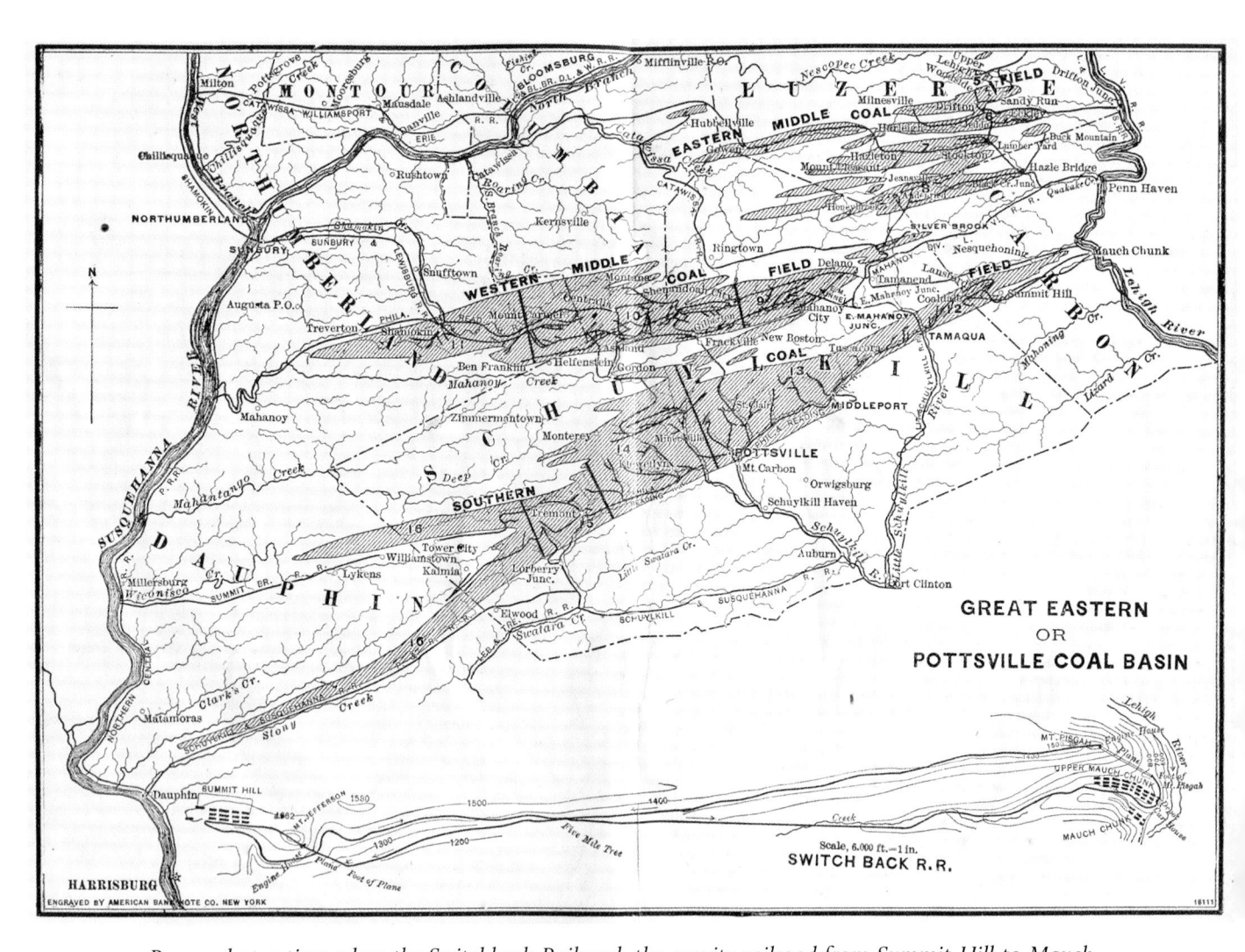

Prepared at a time when the Switchback Railroad, the gravity railroad from Summit Hill to Mauch Chunk, was a popular tourist attraction, this map shows the extent of the Pennsylvania anthracite coal fields. Note that the section from Summit Hill to Mauch Chunk is greatly enlarged at the base of the map, in order to show the features of the Switchback.

The decade 1820 to 1830 saw an enormous expansion of mining along with development of river transportation systems to carry hard coal to major markets. These were the Lehigh Navigation–Delaware Canal, upgraded into a two-way system, the Schuylkill Canal, and the Delaware and Hudson Canal.

(NCM/D&L)

Chapter Two

1820 to 1830
Coal and Canals

"In the Spring of 1827 it was finally concluded we were strong enough to begin & prosecute the ascending Navigat^n. for the prosecution of that great undertaking the C^o. employ'd Canvass White, Esq. as the principal Engineer …"

Josiah White's *History given by himself,* 1832

Pennsylvanians who lived and worked east of the Alleghenies embraced anthracite enthusiastically during the 1820s. Anthracite quickly proved to be its own best advertisement. It was long-burning, clean, gave off very little smoke, and made lots of heat. In the larger cities like Philadelphia, the absence of smoke alone was a strong selling point. Better yet, anthracite was much cheaper than wood for heating. Business historian Alfred D. Chandler records that the head of the Pennsylvania Hospital reported in 1825 that his annual heating costs went from $3,200 to about $2,100 after switching to anthracite.

Many manufacturers found that a ton of anthracite could do the work of 200 bushels of charcoal, which resulted in significant cost savings: a bushel of charcoal, weighing about 32 pounds, was 6 to 8 cents per bushel (thus, 200 bushels cost $16), while anthracite cost $7.50 to $8.00 a ton. Anthracite's high heat and efficiency also made it preferable and cheaper overall than Virginia bituminous coal, which had become available again after the end of the War of 1812.[1]

For more than five years, the Lehigh Coal and Navigation Company, which was formed from the merger of the Lehigh Coal and the Lehigh Navigation companies in 1821, had a virtual monopoly of the Philadelphia anthracite market. The Schuylkill navigation system, which would give Philadelphia access to the coalfields of the upper Schuylkill, did not commence operation until 1825, when it brought 5,306 tons of anthracite to Philadelphia. In contrast, the Lehigh Coal and Navigation Company landed 28,393 tons of anthracite at Philadelphia in 1825.[2]

By the mid-1820s, the initial problems with anthracite had been solved. Multiple types of fireplace grates and stoves were available, which made igniting the coal and using its hot, long-lasting heat safe and easy to use in homes, shops and other businesses, and industry. The only thing holding back the complete triumph of anthracite was getting even more of it to the people who wanted to use it. That was about to change.

The only known depiction of a coal ark going through a "bear-trap" lock is in the Lehigh Coal and Navigation Company's letterhead. This bill of lading dates from 1828, and shows that the company was shipping coal as far as upstate New York by schooner by this time. (NCM/D&L)

The Canals Carry Coal—and a Revolution

A team of six horses or mules could pull four or five tons in a Conestoga wagon driven by one teamster, while two mules or horses could haul a hundred tons floating on a canal boat with a crew of two. With luck, a wagon could cover 15 miles a day, while a canal boat moving at an average of two miles an hour could travel 36 miles during an 18-hour day. These simple comparisons make clear why canals began to replace roads as the principal transportation routes in the 1820s.

Though other canals, most notably the Middlesex Canal in Massachusetts and the Erie Canal, were completed earlier, the importance of the completion of the Lehigh Navigation cannot be overstated. The Lehigh Navigation—the combined canal ditch and slackwater river system that linked Mauch Chunk and Easton—was the first commercially successful transportation route between the anthracite coalfields of northeastern Pennsylvania and Philadelphia. This was the final ingredient needed to make anthracite the most popular fuel in the northeastern United States.

Demand had been created by the introduction of stoves and fireplace grates that could handle the high heat of burning anthracite without melting or cracking. All that was lacking was an adequate supply of anthracite at affordable prices; this was solved by the completion of the Lehigh Navigation. Large quantities—an increase from 9,500 tons to 35,000 in one year, including a much smaller amount

Patent dated 1822 for 240 acres, 140 perches of forest land in Pokono Township (then in Northampton County), bought by Josiah White and Erskine Hazard in 1822 for $4.88, plus $10 in fees.
(NCM/D&L)

carried by the unfinished Schuylkill Navigation—were delivered to Philadelphia for use in the city itself and for export to New England. Chandler states that it was this sudden and significantly large availability of a fuel vastly superior to anything else available on the continent that "provided the fuel that modernized the American iron industry east of the Alleghenies and shifted its location from the isolated hills of central Pennsylvania to the waterways in the eastern part of the state."[3]

In 1821 White and Hazard reorganized their enterprise by combining the Lehigh Coal Company and the Lehigh Navigation Company into the Lehigh Coal and Navigation Company, a firm that would remain in business until 1987. The Lehigh Coal and Navigation Company was formally incorporated in 1822. It also acquired vast forest acreage in northeastern Pennsylvania. This was a prudent move, because the constant need to build coal arks, which made only one downstream trip before being dismantled and sold for lumber in Philadelphia, was contributing to the rapid disappearance of timber resources in the Lehigh Gorge and Pocono regions. Timber was also necessary to shore up the shaft mines that the company was rapidly opening as the easily accessed surface mines were depleted.

The extremes of both depth and current in the wild Lehigh River severely limited the size and capacity of boats carrying the coal. In 1826 the LC&N commissioned Isaac Chapman to map the Lehigh River as the first step in constructing a two-way canal-and-slackwater navigation. The success of the Erie Canal, completed in 1825, showed that American engineering was up to the task of not only making major alterations to the landscape, but also building the lift locks necessary for an ascending and descending system. To oversee their project, White and Hazard hired Canvass White, a brilliant young engineer who had honed his engineering skills in the completion of New York's Erie Canal.

Canvass White

Begun in 1827, the rebuilt Lehigh Navigation system was a civil engineering wonder when it opened on June 26, 1829. Not only was it an unprecedented construction achievement, it was completed far ahead of schedule, and under budget. These feats were accomplished by a combination of ingenious design and what we now call due diligence. Canvass White's experience on the Erie Canal had taught him the wisdom of working with the terrain wherever possible. So some stretches of the canal were simply dug into the hills along the Lehigh River, which effectively made the hills the berm bank (opposite the towpath bank) of the canal. Apparently, the work crews were directed to dig only half the depth needed, and achieve the required depth by piling up the excavated earth to build up the towpath bank high enough to provide the remainder of the depth.

The limestone and shale soils of the Lehigh Valley would have made the canal portion of the Navigation a poor water holder, so Canvass White employed a technique he had seen in British road building. After directing his crews (many of whom were Irish laborers who had survived the Erie Canal construction) to stockpile any deposits of natural clay they discovered, they "puddled"

Canvass White, 1790–1834

Canvass White was born in Whitestown, Oneida County, NY, in 1790. He was plagued by ill health from childhood (possibly by a genetic pulmonary disorder), which kept him from taking on the usual physical chores on his family's farm and grist mill. His father put great emphasis on education, so White attended the well-regarded Fairfield Academy, where he excelled at mathematics, chemistry, astronomy, mineralogy, and surveying. This education laid the foundation for his extraordinary career.

In 1816, White was hired by Benjamin Wright, who, with James Geddes, had been selected by the New York Canal Commissioners to survey and plan a route for the Erie Canal. At the time, there were essentially no American engineers. Wright and Geddes themselves were surveyors, though Wright had acquired some engineering experience as an assistant to English canal engineer William Weston. White, the only assistant surveyor with formal education, became Wright's chief assistant. Then he was selected to spend six months in Britain studying canal and bridge engineering. White returned not only with a deep knowledge of how to build a waterway, but with more accurate surveying tools than were available in the United States, and, most importantly, samples of hydraulic cement, which hardens underwater.

During the years of construction of the Erie Canal, White's engineering skills and problem-solving capabilities moved him into the top rank of the fledgling field of civil engineering. During the 1820s, he was the chief engineer of the Union Canal, where he built the longest tunnel and the largest dam in the U.S. to that time; was chief engineer of the Lehigh Navigation; directed construction of the locks and short canal around the Enfield Falls on the Connecticut River; acted as the consultant on the Farmington Canal; surveyed the route of the Juniata Division of the Pennsylvania Main Line canal and apparently proposed what became the Allegheny Portage; turned down the offer to become chief engineer of that undertaking and all of Pennsylvania; laid out the route of the Camden and Amboy Railroad; and designed and directed construction of the Delaware and Raritan Canal. During all of these undertakings, he trained the next generation of engineers, including John Jervis, Simon Guildford, Sylvester Welch, and William Lehman, all of whom made the transition from canal building to railroads. But his chronic ill health cut his life short, and he died in Florida in December 1834, at the age of only 44.

"White's career demonstrates that canal engineers launched the technological revolution in the United States," wrote Gerald Bastoni. "He helped to marshall resources and knowledge from this country and England, and apply them to achieve a breakthrough that had a profound impact ... Just as importantly, he imparted a confident, self-reliant attitude by his own example that bolstered a rising generation of civil engineers faced with new and challenging problems."[4]

Rebuilding a timber-and-rock dam after the devastating flood of early 1902. This is "Slate Dam" at Laury's Station.

The photo shows the way dams along the two-way Lehigh Navigation were built. Timbers were fastened together to create cribs into which large rocks were placed, and heavy wooden sheathing covered the structure to allow water to sheet-flow across.

Above the dam (photo on left) was the slackwater pool. On the far bank is a stone wall with a guard lock and the locktender's house. All locks were lined with wood inside the chamber, one more reason that vast forest reserves were essential.

The photo below shows the view from downstream as repairs at the White Haven dam are nearing completion in 1908.

(both photos: NCM/D&L)

Mule-drawn boats traveled along the edge of slackwater pools and entered or exited the canal at the guard locks.

The dams and guard locks diverted river water into the canal.

The Delaware Canal was entirely separate from the river and had no dams or guard locks. From the beginning it had problems maintaining an adequate level of water for the boats.

the clay by mixing it with water and then pressing the water out. The puddled clay was then applied to the walls and bottom of the canal prism, and tamped down by allowing herds of local sheep to run up and down the canal bed. This minimized the loss of water through leakage, a problem that plagued other nineteenth-century canals. Muskrats tunneling in the canal banks caused leaks and even occasional breaches. In an attempt to control the problem, the canal company offered a bounty of ten cents per dead muskrat.

Another time and money saver was incorporating the Lehigh River itself into the new navigation. Eight timber-and-rock dams converted sections of the river into slackwater pools, which combined to make up 10 miles of the Lehigh Navigation's 46-mile route between Mauch Chunk and Easton. The slackwater pools provided a steady source of water for the ditch sections of the canal, which were 5 feet deep, 60 feet wide at the top, and 45 feet wide at the bottom. Boats passed from the slackwater pools into the canal sections through eight guard locks.

The Lehigh Navigation had to overcome a 355-foot difference in elevation between Mauch Chunk and Easton. This was accomplished using 48 lift locks of unprecedented sizes. With the exception of locks numbers one to four, which were built immediately downstream from Mauch Chunk, the locks were 22 feet wide and 100 feet long; the four near Mauch Chunk were a staggering 30 feet by 130 feet. At its completion in 1829 the Lehigh Navigation was the largest-capacity towpath canal in America, capable of handling mule-drawn boats that carried over 200 tons of anthracite. Unfortunately, this waterway was never able to reach its full potential because the connecting Morris and Delaware canals were built to much smaller dimensions.[5]

The growing town of Mauch Chunk, from Eli Bowen's 1852 Pictorial Sketch Book. *Small cars can be seen descending or ascending all the tracks of the Gravity Railroad between Summit Hill and the riverside coal wharfs in the right foreground, where canal boats were loaded for their 46-mile trip to Easton.*

Not seen in this engraving is the weigh lock, which was half a mile south of Lock No. 1 at Mauch Chunk. Loaded boats entered the weigh lock (seen on the front cover), which was then drained to rest the boat on the scale. The lock was refilled, and the boats continued south. All boats were weighed empty at the start of the season, then weighed when loaded at the start of each trip. The weight of the coal was determined by subtracting the recorded weight of the boat. Lock No. 2 was alongside, to allow returning boats to bypass the weigh lock.

(Private collection)

Bethlehem / CITIES & TOWNS

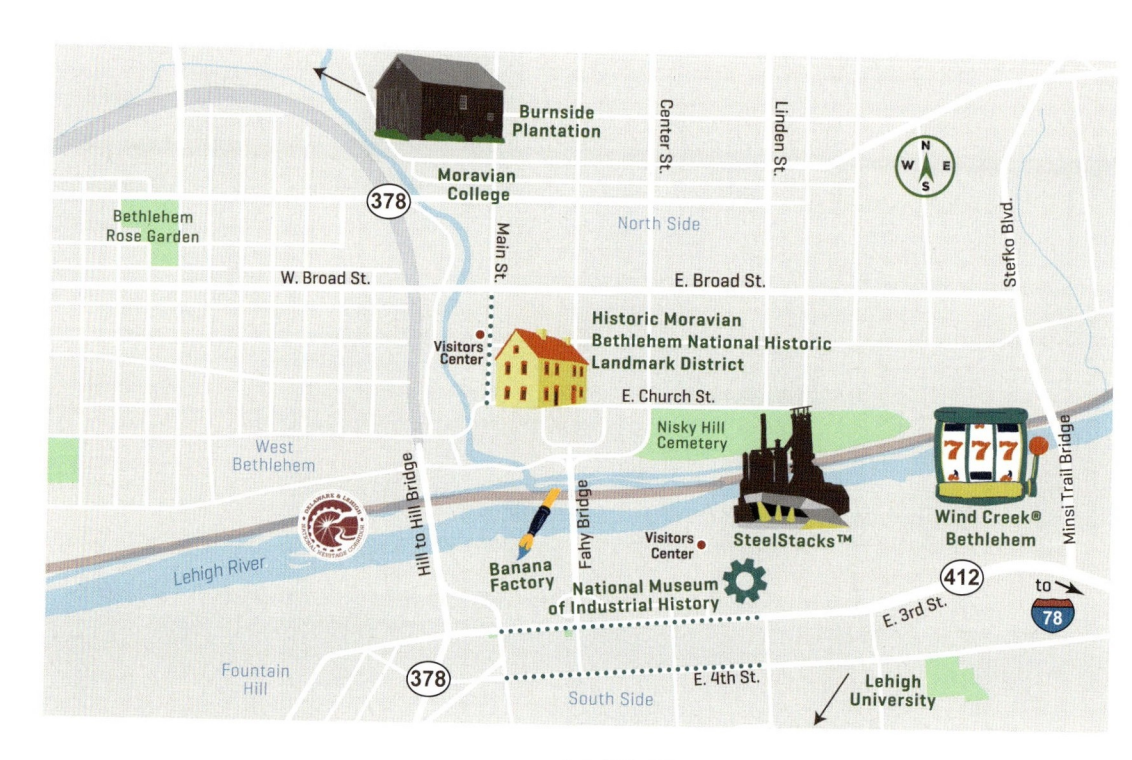

BETHLEHEM

Divided by the Lehigh River, the city of Bethlehem blends pre-Colonial history with modern arts, dining, and more. Explore the city's North Side, which is home to the various properties of Historic Bethlehem Museums & Sites (learn more about their initiatives on page 22), the luxurious Historic Hotel Bethlehem, which was named as one of the top five historic hotels in the country by *USA Today 10Best*, and Main Street which is lined with picturesque shops and dining options. Cross the river to the city's South Side Arts District to dive into an emerging arts & cultural scene, with ArtsQuest™ and the SteelStacks™ campus at its core. The backdrop of the blast furnaces of the former Bethlehem Steel set the stage for some of the most innovative event programming on the East Coast, including Musikfest®, one of the largest and most diverse music festivals in the nation with over 500 free shows across 10 days. »

2020 **Discover Lehigh Valley®** Visitors Guide | 11

CITIES & TOWNS / Bethlehem

THINGS TO DO IN BETHLEHEM

Hoover Mason Trestle (at right) runs along the SteelStacks™

HOOVER MASON TRESTLE

The Hoover Mason Trestle in Lehigh Valley connected the ore yards, once located near where Wind Creek® Bethlehem stands, to the blast furnaces at the heart of the steel plant. Today, visitors can walk the elevated pathway to explore the former site of Bethlehem Steel and discover its history.

711 First St., Bethlehem | HooverMason.com

>> For more on the Hoover Mason Trestle, check out our video at YouTube.com/DSCVRLehighValley

KEMERER MUSEUM

Explore one of the only museums on the East Coast dedicated to the decorative arts. The Kemerer Museum houses an impressive collection of historic dollhouses, furniture, china, and more.

427 N. New St., Bethlehem
HistoricBethlehem.org

Kemerer Museum

WIND CREEK® BETHLEHEM

Feeling lucky? Try your hand on the floor of the casino (home to slots, table gaming, and stadium gaming) at Wind Creek® Bethlehem. The venue boasts restaurants by Emeril Lagasse and Buddy Valastro, on-site shopping, a concert venue, and a full-service hotel.

77 Wind Creek Blvd., Bethlehem
WindCreekBethlehem.com

TOUCHSTONE THEATRE

Rooted in the traditions of street performance and physical theater, Touchstone Theatre produces original musicals and plays, new adaptations of classic texts, and community-based projects. Each spring, the theatre hosts a Young Playwright's Festival to showcase aspiring new artists. 2020 celebrates the 15th anniversary of this program.

321 E. Fourth St., Bethlehem
Touchstone.org >> *Coupon on pg. 61*

BETHLEHEM FOODIE FAVORITES

@prongs_down

TAPAS ON MAIN
With a prime location on Main Street, Tapas on Main is your go-to for Spanish fare. Order a selection of small plates to share—and don't forget the sangria! Favorites include classic red, blood orange, classic white, and rotating seasonal options.

500 Main St., Bethlehem
TapasOnMain.com

JENNY'S KUALI
Transport your taste buds at this SouthSide Bethlehem establishment. Choose from a variety of noodle dishes, dumplings, and more. The menu at this BYOB staple also offers an impressive selection of vegetarian options.

102 E. Fourth St., Bethlehem
JennysKuali.com

theMINT GASTROPUB
From award-winning mac & cheese to one of the best craft beer selections around, theMINT showcases unique culinary offerings in West Bethlehem. Don't miss the $25 Burger, made with grass-fed wagyu beef and pork belly, and always opt for a side of tots!

1223 W. Broad St., Bethlehem
BethlehemMint.com

CHECK OUT MORE DINING RECOMMENDATIONS PG. 41

12 | DiscoverLehighValley.com

A remaining section of the original coal road in Mauch Chunk, built in 1818–1819 to bring coal from the mines at Summit Hill to the Lehigh River for transshipment.

This section of the road became known as Trap Alley, because it was here that Josiah White had the shop where he developed the "bear trap" dam. The story is that when curious people inquired about what was going on in the shop, White would tell them he was building bear traps. Hence the name bear trap dams.

(Courtesy of Vince Hydro)

The Gravity Railroad

It is astonishing to consider that at the same time White and Hazard were wrestling with the challenges of water transportation on the Lehigh River, and ultimately building the two-way Lehigh Navigation, they were also building the first truly successful railroad in America.

With fast-growing demand for their coal, and construction of a vastly more efficient system of transporting to market underway, White and Hazard knew they had to address the issue of getting their coal from the mine on top of a mountain to the boats waiting at its foot.

The chronicle of the Summit Hill to Mauch Chunk Gravity Railroad began with the survey of a road from Summit Hill to Mauch Chunk by Josiah White and Erskine Hazard during 1818. The route of this nine-mile road was laid out with the intent that it could be later converted into a railroad. The road was used by the wagons that hauled anthracite from the mine of Summit Hill to the navigation loading docks at Mauch Chunk. By 1825–1826, shipments of anthracite had begun to overtax the road's hauling capacity and its conversion to a railroad began.

Work began on the Summit Hill to Mauch Chunk Gravity Railroad in January, 1827. Built under Josiah White's direction, this line used the western portion of the already existing nine-mile coal-hauling road, but its eastern portion was placed at a higher elevation in order to minimize the impacts of noise, coal dirt, and the potential danger of derailments on the new settlement of Mauch Chunk. When it was completed in May of 1827, the railroad included more than twelve miles of track, a nine-mile train route between Summit Hill and Mauch Chunk, and almost four miles of branch roads leading to the coal mines. The railroad was built from twenty-foot-long wooden rails that were four inches wide by five inches thick and faced with iron straps.

When it was completed, the Gravity Railroad was the third railroad to operate in North America, the first to be more than five miles in length, and the first to carry large volumes of material. Anthracite was loaded into coal cars at Summit Hill, then the loaded cars coasted by gravity

By all acounts, the mules that rode down the Gravity Railroad enjoyed the trip. They traveled three abreast, troughs of feed in front of them, behind the coal cars and cars filled with tourists. Once the coal cars were emptied at the foot of the inclined plane the mules hauled the cars back to Summit Hill and the mines.

Industrial tourism was making the anthracite area popular as a destination as early as the 1820s. Periodicals of the time told their readers of the exciting discoveries and innovations taking place there. Tourists, men and women, began coming to Mauch Chunk to breathe the fresh air, walk in the mountains, see the famous mines, and enjoy a thrilling experiance riding down the gravity railroad, often described as America's first roller coaster.

This sketch was prepared for and published by the Mumford brothers, who leased the railroad in 1880 and ran it for tourists for several years.

(NCM/D&L)

to Mauch Chunk, where they were unloaded and the empty cars pulled back to Summit Hill by mule teams that had been carried along on the descending trip in their own specially built cars. The system worked efficiently; Summit Hill was 975 feet higher in elevation than Mauch Chunk, and the slope of the descending track was more than 100 feet per mile.

The Summit Hill to Mauch Chunk Gravity Railroad proved to be a great success. During its first year of operation it transported almost 28,000 tons of anthracite. Each loaded coal train included not only fourteen cars carrying two tons each, but also the cars carrying the mule teams. Great care was taken to ensure the fitness of the mules, as can be seen from the following passage taken from the Annual Report of the Lehigh Coal and Navigation Company for 1828:

> Since we have reduced the velocity of travelling from 12 to 15 miles an hour down to 5 to 7 miles an hour, our horses and mules, which in the former rate got and kept sick, the latter continue healthy, notwithstanding their regular daily work is 35 to 45 miles per day; and so strong is their attachment to riding down, that in one instance, when they were sent up with the coal waggons, without their mule waggons, the hands could not drive them down, and were under the necessity of drawing up their waggons for the animals to ride in.[6]

The Gravity Railroad, later renamed the Switchback Railroad, remained in operation for more than a century. In its later incarnation as a tourist attraction, the Switchback is considered by modern amusement park historians to be a progenitor of the roller coaster.

It was not until 1828 that a competing anthracite transportation system could successfully challenge the Lehigh Coal and Navigation Company's domination of the Philadelphia market. In that year the Schuylkill Navigation, which enjoyed the advantage of having a shorter distance from mine to market, began to surpass the LC&N's annual tonnage of anthracite delivered to Philadelphia. Rather than cutting into the LC&N's profits, however, the wider availability and increased efficiency of transporting anthracite served only to increase demand for it.

The Wyoming Coal Fields

The success of the Lehigh Coal and Navigation Company attracted the attention of Wyoming Valley coal entrepreneurs. Although it is estimated that by 1827 coal mines of the Wyoming Valley were shipping approximately 25,000 tons of anthracite annually down the Susquehanna River to the markets of central Pennsylvania and the Chesapeake Bay region, the Wyoming Valley lacked the all-important direct water-transportation route to the Philadelphia market. So complete was the Lehigh Coal and Navigation Company's control of the Philadelphia market that when Philadelphia merchants Charles, Maurice, William, and John Wurts attempted to develop their coal lands in the upper Lackawanna Valley and ship anthracite from there to Philadelphia in 1820–1821, they found that the Lehigh Coal and Navigation Company's lower costs and shorter route between the anthracite coalfields and Philadelphia had made it virtually impossible for them to sell their coal in the Quaker City.[7]

Canals and river navigation systems were the only ways to transport bulky products to market in large volume at an affordable cost before long-haul railroads were constructed in the second half of the nineteenth century. The Wyoming Valley coal operators focused their attention on the Susquehanna and its improvement, which would eventually result in the building of the North Branch Canal System as a part of the Pennsylvania state-built canal system.

The Wurts brothers decided their only hope would be to supply the New York City market. They engaged Benjamin Wright, who had engineered the Erie Canal, to survey a canal route between their mine and the Hudson River. They also shipped some coal to New York, where they demonstrated its usefulness to potential investors. The Delaware and Hudson Canal Company was organized on March 8, 1825, and soon became the third great Pennsylvania anthracite canal. This led to the development of the anthracite industry in the upper Lackawanna Valley, and the founding of Carbondale and Honesdale.

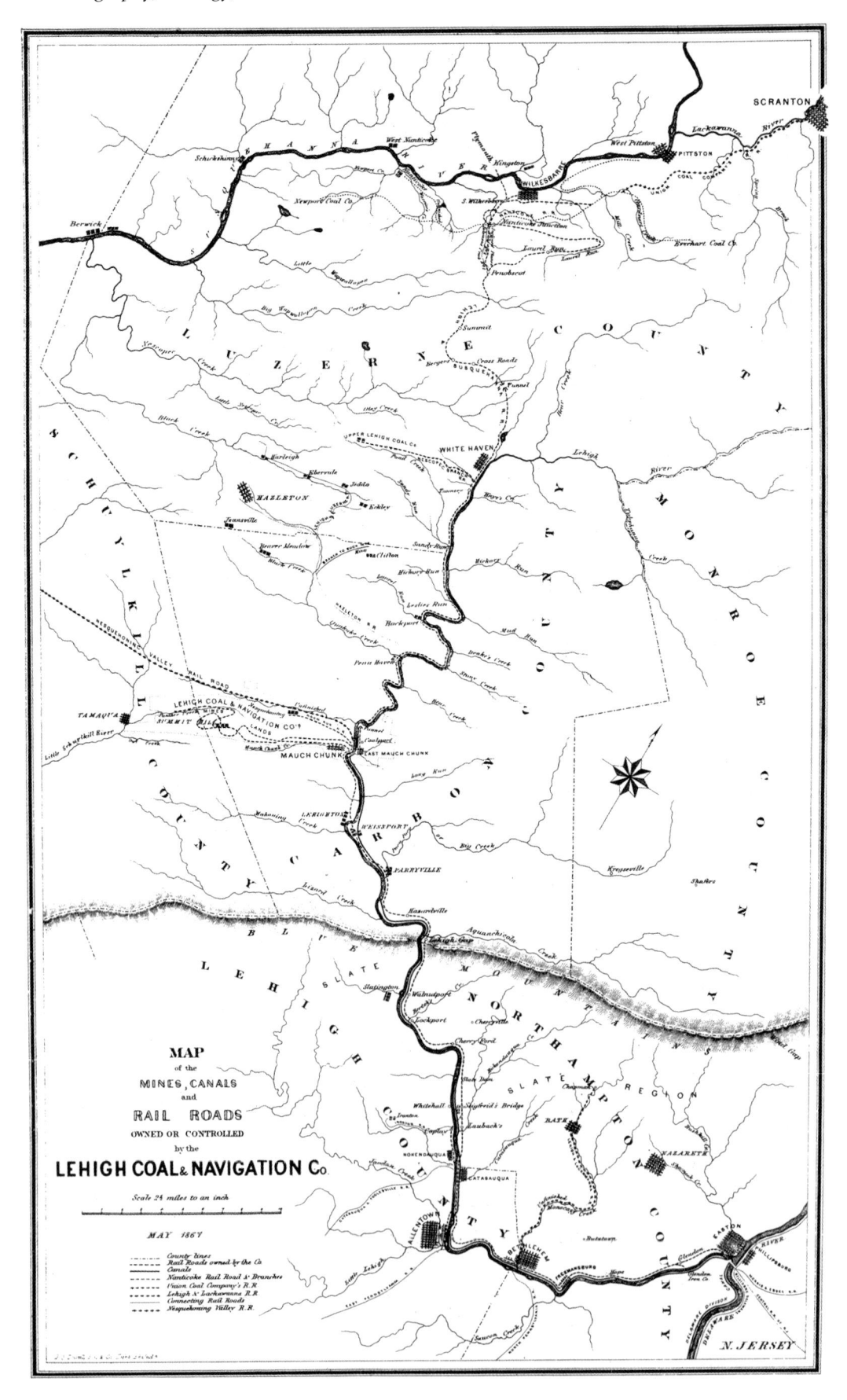

MAP
of the
MINES, CANALS
and
RAIL ROADS
OWNED OR CONTROLLED
by the
LEHIGH COAL & NAVIGATION Co.

Scale 2½ miles to an inch

MAY 1867

Chapter Three

1830 TO 1840
CONNECTING MINE TO MARKET

*"And we have now lived to see the day that the public & Engineers is
<u>as much opposed to small</u> Canals & <u>Small</u> Locks as they were
at the beginning of our Canal, Opposed to <u>Large</u> Canals & Locks."*

Josiah White's *History given by himself, 1832*

With the Lehigh Coal and Navigation Company's mining and transportation systems moving like clockwork, Josiah White could turn his attention to the rebuilding of the badly constructed Delaware Division Canal. Even as devout a Quaker as White must have had some difficulty suppressing the urge to point out that if the Pennsylvania legislature had not chosen ten years earlier to build a too-small canal system on the cheap, they would not now be turning to him, grudgingly, to salvage a botched, costly job.

As early as 1823, White and Hazard had petitioned the Pennsylvania legislature for the right to improve navigation on the Delaware River. Their initial plan called for the construction of a series of dams and locks that would convert the Delaware River between Easton and Philadelphia into a huge ship canal. This waterway would allow steam tugs towing sailing schooners to ascend and descend the Delaware–Lehigh river system.

White and Hazard's dream was never realized because of the intervention of politicians and rival entrepreneurs. State Senator Peter Ihrie and Bucks County entrepreneur and political power Samuel Ingham proposed legislation in 1826 that would allow the Commonwealth of Pennsylvania to build a smaller canal from Easton to Bristol that would be entirely separate from the Delaware River. Ihrie despised White and Hazard as monopolists who controlled the commerce of the Lehigh Valley, while Ingham hoped to use this newly proposed waterway as part of a transportation route to coal lands that he owned near the present site of Hazleton.[1]

Construction of the Delaware Canal

On October 27, 1827, construction began on the Delaware Division Canal. It was called this because it was a part of the Pennsylvania canal system, even though it did not connect with any other part of this massive public works. By 1829 over 53 miles of the planned total length of 60

Circa 1890 view of the Forks of the Delaware with Williamsport, known popularly as Snufftown, and Mount Ida in the center. Coal yards and businesses lined the banks. Hundreds of people lived here during the canal era, particularly in the mid-nineteenth century, with stores, maintenance shops, hotels and taverns serving the boatmen and canal workers. A series of floods seriously damaged or destroyed many of the buildings remaining after the peak of activity on the canals, and during the twentieth century the area declined rapidly. (NCM/D&L)

miles had been completed. The canal's northern end in Northampton County was just to the south and east of the mouth of the Lehigh River in an area that was variously known as Shaefferstown, Snufftown, and South Easton.

Water for the Delaware Division came from the Lehigh River. The dam that the Lehigh Coal and Navigation Company had built across the mouth of the Lehigh River at Easton (which created the slackwater pool that reached back to the Lehigh Canal's outlet at South Easton) became the means of supplying water for the Delaware Canal. A guard lock and feeder near the Easton dam linked the Lehigh Navigation and the Delaware Canal.

Bristol, on the Delaware River in Bucks County, was the logical southern terminus of the Delaware Division. The third-oldest town in Pennsylvania, Bristol had been a transportation hub since Samuel Clift established a ferry across the Delaware River there early in the 1680s. Only a few years later, in 1689, a road known as the King's Highway passed through Bristol, linking Morrisville to Philadelphia. Because it is located below the Delaware's fall line and hence was accessible to coastal and even ocean-going ships, Bristol had boasted shipyards and a flourishing wharf and warehouse trade since the eighteenth century. Water power harnessed from Mill (later Otter) Creek and Mill Pond (which was incorporated into the canal) as early 1701 allowed for the production of flour and sawn lumber.

Josiah White and Delaware Canal assistant engineer Lewis Coryell installed waterwheels to pump water from the river into the canal here at Union Mills, south of New Hope. The system was completed in 1832 with a feeder dam in the river to direct water into the channel that led to the waterwheels, and a chute in the center of the river for the many rafts of logs coming down the Delaware. The water wheels were rebuilt in 1880. They pumped 3,500 cubic feet of water per minute into the canal until they ceased operation in 1923. (NCM/D&L)

Too Small and Badly Built

By 1830 the Delaware Division Canal was declared to be finished—despite the fact that the water diverted by the dam at Easton was not enough to fill the full length of the canal. This alone posed a significant problem, but because the Canal Commissioners had chosen the fastest and cheapest design and building option every time, it quickly became apparent that shoddy construction and poor engineering meant the canal would not hold any water at all. Banks breached and leaked, locks fell in, and aqueducts collapsed. The Delaware Division Canal soon became a stagnant, unusable ditch.

Attempts to repair it failed miserably. It was only with great difficulty that a boat loaded with anthracite finally passed through it to Bristol in 1832. In desperation, the Commonwealth appointed Josiah White, assisted by Louis S. Coryell, to supervise what amounted to rebuilding the canal. Working together, White and Coryell re-engineered the entire canal; they solved the problem of getting it fully watered by designing and building a waterwheel pumping system to feed water from the Delaware River into the canal at New Hope. When completed, this pumping system supplied the Delaware Division with 3,500 cubic feet of water per minute. It functioned successfully until 1923. By the end of 1832, White's able management and engineering skills had put the Delaware Division Canal into working order. With the completion of a weigh lock at Easton in 1833, the canal was ready to be placed into commercial operation. The first year with a full boating season was 1834.

The Delaware Division Canal had a much smaller carrying capacity than did the Lehigh Navigation. The original 23 lift locks of the Delaware, a mere 11 feet in width and 95 feet in length, were only half the width of the lift locks of the Lehigh Navigation. This small size of lock, adopted on the recommendation of engineer and architect William Strickland, infuriated White and Hazard. Strickland had been sent to England in 1825 by the Pennsylvania Society for the Promotion of Improvements, where he observed the long, narrow locks that had been in use for many years. His report noted that such locks were expeditiously and conveniently filled, were cheaper,

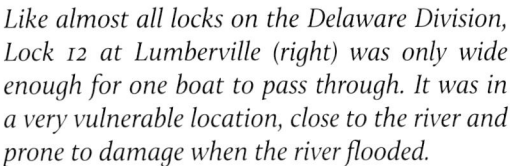

Like almost all locks on the Delaware Division, Lock 12 at Lumberville (right) was only wide enough for one boat to pass through. It was in a very vulnerable location, close to the river and prone to damage when the river flooded.

The photo above left shows LC&N boats tossed about following the flood of 1903. This one is a "section" boat, with two sections hinged together in the middle. This facilitated unloading coal, as the two sections could be unloaded essentially as two separate boats.

The watersheds through which the Lehigh and the Delaware flowed were very different, as were the floods on each. The flood of 1902 was one of the worst ever on the Lehigh, and the flood of 1903 was far worse on the Delaware. Damage differed also: the Lehigh passed through more areas that became developed with industry and warehouses close to the river, while the Delaware remained much more rural.

(NCM/D&L)

required little water, and saved time. White and Hazard had constructed the Lehigh Navigation to allow the passage of boats carrying 200 tons of cargo; the small size of the lift locks on the connecting Morris and Delaware canals meant that the Lehigh Navigation was in effect limited to half of its designed capacity.[2]

Despite this, by 1836 the Delaware Division had the distinction of being the only part of the Pennsylvania Canal system that was making money. This was, of course, because of the ever-growing volume of anthracite coming down the Lehigh Navigation, headed for Philadelphia. This made the problem of the Delaware Division's smaller locks ever more vexing. But not only was the Canal Commission reluctant to spend the money to enlarge the locks, there was also the problem of how to supply the greater amount of water in the canal that larger locks would require.

The issue became moot when two destructive floods, one in January 1839, and the second on April 8 the same year, struck the Lehigh and the Delaware. The south end of the dam at the mouth of the Lehigh at Easton was destroyed, but rebuilt quickly thanks to "a heavy force" of workers raised and supported largely by the citizens of Easton: "the breach was attacked with vigor" and closed by mid-February.[3] Many of the Delaware Division's aqueducts were severely damaged or destroyed. The wooden locks, already wearing out after only nine years in operation, also sustained major damage.[4] The Canal Commissioners decided to use the destruction as an opportunity to enlarge the locks, but with the Commonwealth in severe financial straits due to the Panic of 1837, much of the cost of rebuilding and repairing the canal was borne by the Lehigh Coal and Navigation Company and the Beaver Meadow Company, both of which had a major interest in resuming coal shipments as soon as possible.[5] Very few of the locks were rebuilt then, or ever, resulting in slowdowns when more than one boat needed to lock through.

Repairs continued through the boating season, the start of which was delayed until May 14, 1839. Despite the late start, and a two-week shutdown in late August due to a breach in the towpath three and a half miles south of Easton, tolls collected on the Delaware Division were $26,000 higher than the previous year.

The next year—1840—was to bring the most momentous change of all to the counties in the valleys of the Lehigh and the Delaware.

Construction of the Lehigh Navigation's Upper Grand Section

Once he had overhauled the Delaware Canal and made it into a workable, albeit imperfect, link between his canal and Philadelphia, White turned the focus of his energy and ingenuity on yet another major engineering challenge: the northward extension of the Lehigh Navigation from Mauch Chunk through the Lehigh River Gorge. Building this waterway would allow the Lehigh Coal and Navigation Company to tap the expanding coal mines of both the Eastern Middle Coal Field and the Wyoming Valley. Of equal legal importance, it fulfilled a provision in the Lehigh Coal and Navigation Company's state charter that called for a connection with the Susquehanna River.

The design and construction of this canal was a daunting task because the elevation of the Lehigh River fell almost 600 feet in the 26 miles between the site that became White Haven and Mauch Chunk. To oversee construction of the waterway the Lehigh Coal and Navigation Company hired a young engineer, Edwin A. Douglas. Douglas had learned canal engineering working with Canvass White on the construction of the Erie and the Delaware & Raritan canals, making him one of the most experienced canal builders in the U.S. in the 1830s. He was not afraid to take on the challenge of taming the turbulent upper section of the Lehigh River. Confident that he could execute Josiah White's plans for a series of high dams and locks on the turbulent Lehigh River between Mauch Chunk and White Haven, he requested the handsome salary of $4,000 a year. For the LC&N, it was money well spent.

Edwin A. Douglas

Work began on the extension of the Lehigh Navigation in 1835. Overcoming the 600-foot elevation change involved building 29 massive timber-and-rock crib dams and 29 locks in fewer than 26 miles of the Lehigh River. Some of the locks were huge buttressed stone structures that were capable of raising and lowering boats an unprecedented 30 feet.

Douglas greatly speeded up locking time by developing the drop gate, which was installed at the upstream end of the locks of the extension. Horizontally hinged drop gates enabled boats to enter and leave locks quickly; this innovation was so successful that drop gates were later installed on the lower portion of the Lehigh Navigation and later the design was adopted for use on the Morris, Delaware and Raritan, and Delaware and Hudson canals. Drop gates were also installed on a few selected locks of the Delaware Division Canal that were later enlarged by the Lehigh Coal and Navigation Company.[6]

Drop gates were not Douglas's only innovation. It quickly became apparent that the traditional method of opening and closing mitre gates on locks by means of long, heavy balance beams was not practical with such deep locks in the confined spaces of the Lehigh Gorge. So Douglas invented a rack-and-pinion gear system for moving lock gates; this not only took up much less room (a so-called "dog house" is about three feet square) but was vastly more efficient. The gear system made it possible for locks to be operated by no more than one or two people, who could even be women or older children. It is important to note that all of this canal construction, which was taking place not only in Pennsylvania, but also in New York, New Jersey, Ohio, Indiana, and Maryland, was carried out entirely by human and animal muscle, using basic tools and simple machines such as the lever, pulley, inclined plane, and wheel and axle. The combination of drop gates and "dog houses"—all mechanical devices that were never operated by any other means than human muscle—greatly reduced the time it took for boats to traverse locks. The design was adopted by virtually all the canals in the Delaware Valley, and was used for the entire existence of the Lehigh Navigation and the Delaware Canals.

The northward extension of the Lehigh Navigation was called the Upper Grand Section. The northern terminus rapidly developed into the town of White Haven, which was connected to Stoddartsville in 1838 by a further extension north along the Lehigh River. With its bear-trap locks, this extension was capable only of descending navigation and was used mainly for the shipment of lumber.

THE LEHIGH AT TANNERY.

The dam at Tannery, about a mile below the terminus of the Upper Grand Division, was 34 feet high, and the lock 30 feet deep. The dam and lock suffered much less damage in the flood of 1862 than the Navigation's structures downstream, which were completely swept away and not rebuilt. Seen here in an engraving in a Lehigh Valley Railroad guidebook from 1873, the dam and lock were finally destroyed in early twentieth-century floods.

(NCM/D&L)

Railroad Building Begins

Just at the time that canals reigned supreme as coal-movers, railroads—in some cases built by the canal companies themselves—began their incursion into the anthracite industry. Though many of these early steam railroads acted only as feeders to the canals during the 1830s, the engineering, industrial, and economic foundations they laid in that decade led directly to the explosion of American industry, transportation, communication and finance that occurred in the 1840s.

The Beaver Meadow Railroad

The first steam railroad to operate in the five-county area that now comprises the National Heritage Corridor was the Beaver Meadow Railroad. This was the second steam railroad to operate in Pennsylvania's anthracite coal regions; it was preceded only by the Little Schuylkill Railroad, which began operation in 1833, linking the coal lands near Tamaqua with the Schuylkill Navigation at Port Clinton. Construction on the Beaver Meadow only began that year.

The railroad was headed by Samuel Ingham, the politician and entrepreneur from Solebury, Bucks County, who had opposed White and Hazard's plans for the Delaware River, and he found himself at loggerheads with them again. The chief and assistant engineers were Canvass White and Ario Pardee. Under Canvass White's direction a route for the railroad was surveyed that ran south and east from Beaver Meadow along the Beaver and Hazel creeks to a site that would become the town of Weatherly on the Black Creek. From Weatherly the route of the railroad ran along the Quakake Creek to a site that was to become known as Penn Haven, then southward along the Lehigh River.

Ario Pardee, circa 1870.

The route along the Lehigh brought the Beaver Meadow Railroad into conflict with the powerful Lehigh Coal and Navigation Company, which was in the process of extending its waterway northward to a site that would become White Haven. Armed conflict between the construction crews of the two concerns almost erupted, but the superior power of the Lehigh Coal and Navigation Company prevailed and the Beaver Meadow Railroad was forced, at great expense, to alter its route along the Lehigh. Since the Beaver Meadow's management also objected to what it considered very high tolls charged by the Lehigh Coal and Navigation Company for the shipment of coal on its waterway, they decided to extend the route of the railroad southward to Easton. Tracks—wooden rails covered by iron straps, laid on the road bed—were laid as far as Parryville for a total distance of twenty miles before a compromise was reached, making this location the transfer point between the Beaver Meadow's coal cars and the Lehigh Coal and Navigation Company's boats.

By 1836 the Beaver Meadow Railroad had begun regular operation with Ario Pardee serving as its superintendent and Hopkin Thomas its master mechanic. Although Pardee would soon trans-

Ario Pardee, 1810–1892

Ariovistus Pardee grew up on a farm near Lebanon Springs, New York. Essentially self-taught except for two years in his teens when he was tutored by a Presbyterian minister, Pardee nevertheless was hired by E.A. Douglas, a family friend, and Canvass White to help survey and engineer the Delaware and Raritan Canal in 1829. Two years later, he was sent to survey a railroad to carry coal from the new mines at Beaver Meadows to the Lehigh Navigation's docks in Mauch Chunk. He ended up supervising the entire construction and start of the Beaver Meadow Railroad, followed by construction of another rail line that connected the mines near Hazleton to the Beaver Meadow at Weatherly.

In 1840, he went into anthracite mining with two partners, Robert Miner and William Hunt; Hunt was replaced two years later by J. Gillingham Fell. But the land where the firm's mines were located was all owned only by Pardee. In *The Kingdom of Coal*, Donald Miller and Richard Sharpless recount another possibly apocryphal story of anthracite discovery: While he was surveying for the railroads, it was said, Pardee had seen deer scratching up black rocks just under the surface of the ground, and made quiet haste to acquire the lands.

True or not, Pardee soon became the largest shipper of anthracite in Pennsylvania, with mining operations at Hazleton, Cranberry, Sugarloaf, Crystal Ridge, Jeddo, Highland, Lattimer, Hollywood, and Mount Pleasant. By 1863, his coal, lumber, and iron holdings, in Pennsylvania and elsewhere, all forged with his single-minded focus on his businesses, had made him a millionaire.

Unlike some of his fellow coal magnates, Pardee made his home among his holdings, and became largely responsible for the transformation of Hazleton from a "patch" to a thriving town. He helped build the population by fathering 13 children, though one son and his first wife died in childbirth in 1847, and two sons by his second wife died in infancy. A fervent Presbyterian, Pardee founded Hazleton's first church, a company store that became a general public emporium, and public schools. He was passionately interested in promoting the technology and use of anthracite, and endowed a school of science at Lafayette College that focused on mine engineering.

Like many industrialists of his time, particularly those who lived among his workers, Pardee took a benevolent yet authoritarian attitude toward them. Most believed that their superior status gave them both responsibilities toward their workers as well as near-absolute authority over them. Though willing to deal directly with workers to resolve disputes, Pardee and his contemporaries vehemently, even violently, opposed unions.

Nevertheless, Pardee's funeral in Hazleton in 1892 was attended by thousands of people, who came by the trainload, a local newspaper reported at the time. "Millionaires jostled elbows with slate pickers, Hungarian women stood by exponents of fashion," the newspaper reported. "It was a wonderful sight." [7]

Pardee is buried in the Vine Street cemetery in Hazleton.

fer the focus of his activities to the newly formed Hazleton Railroad and eventually become a major anthracite coal operator, Thomas remained with the Beaver Meadow and did much to insure its successful operation.

The Beaver Meadow Railroad quickly proved to be a commercial success; in 1838 it shipped over 4,442 tons of anthracite to its canal loading docks at Parryville. Unfortunately, its prosperity was short-lived. In January of 1841 a terrible storm and accompanying flash flood ravaged the upper Lehigh River Valley, damaging not only the Lehigh Navigation but also the Beaver Meadow Railroad. All of the line's bridges between Weatherly and Parryville were washed away while the roadbed itself was almost totally destroyed between Mauch Chunk and Parryville. Unable to finance the complete repair of the entire system, the company's management authorized a shortening of the line to Mauch Chunk, where new loading facilities were built on the canal.

Hopkin Thomas

The railroad's master mechanic, Hopkin Thomas (1793–1878), was from a generation of skilled and experienced Welsh mechanical engineers who emigrated to the United States and drove much of its early industrial development. His steam engine expertise brought him to Philadelphia's Baldwin Locomotive Works in 1834, and then to Garret and Eastwick, inventors of the first anthracite-powered locomotives. Thomas delivered one to the Beaver Meadow Railroad in 1836, and was promptly offered the post of chief engineer. Thomas patented a series of improvements to locomotive engines, wheels, and axles. In 1853, his childhood friend David Thomas hired him to construct the Crane Iron Railroad and direct the works' foundry and machine, blacksmith, and pattern shops.

Hopkin Thomas retired as master mechanic of the Crane at the age of 80 in 1874. His greatest contribution was the many men he trained, among them Captain Bill Jones, Andrew Carnegie's right-hand man, Bethlehem Iron's Owen Leibert, and Phillip Hoffecker and John Kinsey of the Lehigh Valley Railroad.

The Lehigh and Susquehanna Railroad

Though the Lehigh Coal and Navigation Company was well established and thriving, there was still one unfulfilled clause in its charter with the Commonwealth of Pennsylvania—establishment of a link between the Lehigh and Susquehanna rivers. Since it was not feasible to construct a canal over the almost 2,000-foot Penobscot Mountain, which separates the Lehigh River Valley from the Wyoming Valley, a railroad, the Lehigh and Susquehanna (L&S), was constructed instead.

Edwin A. Douglas, the genius behind the Upper Grand Section, agreed to take on the project, and began surveying shortly after the legislature approved it in early 1837. Building this railroad was a remarkable achievement: it involved digging a 1,742-foot tunnel between White Haven and Solomon Gap, and a series of three double-tracked inclined planes from Solomon Gap to Ashley in the Wyoming Valley. (The location of the Planes can be found near the top of the map on page 22, between Penobscot and Nanticoke Junction.)

The Ashley Planes originally operated by gravity, but in 1847, four years after it opened, steam engines and wire-rope cables were installed to haul cars of Wyoming anthracite over the mountain. At the south foot of the planes, the cars were hauled via the Lehigh and Susquehanna Railroad to White Haven, where the coal was transferred to LC&N boats. The completion of the L&S's first phase in 1843 provided the first commercially successful coal transportation route between Philadelphia and the anthracite mines of northeastern Pennsylvania's upper coalfields.[8]

The Wyoming Valley Anthracite Industry

While the Lehigh Coal and Navigation Company was taking a pre-eminent role in the development of the anthracite coal-mining industry, developments were proceeding at a slower pace in the Wyoming Valley. Until the completion of the Lehigh Coal and Navigation Company's Upper Grand Section of the Lehigh Navigation and the Lehigh and Susquehanna Railroad between 1838 and 1844, Wyoming Valley mining companies had no direct access by water transportation to the Philadelphia market. Instead, they were forced to rely largely on the Susquehanna River route to the much smaller markets of central Pennsylvania and the Chesapeake Bay regions. As a result, Wyoming Valley politicians and entrepreneurs, anxious to bring about improvements in the navigation of the Susquehanna River, took a prominent part in the efforts that took place between 1825 and 1828 to develop the Pennsylvania Public Works, a state-constructed and -operated canal system. One of these public works was the North Branch Canal, completed by 1830, which, through its connections to the Susquehanna Division Canal and later the Susquehanna and Tidewater Canal, greatly improved the transportation link between the Wyoming Valley and the markets of central Pennsylvania and the Chesapeake regions.[9]

Despite the completion of the North Branch Canal, however, sales of Wyoming anthracite lagged far behind the Lehigh and Schuylkill regions until the opening of the Lehigh Coal and Navigation Company's route from the Wyoming valley to Philadelphia in the early 1840s. The following figures for anthracite production of the Lehigh, Schuylkill, and Wyoming regions for the years 1834–1837 illustrate the greater dominance of the Schuylkill and Lehigh fields, directly attributable to the direct access both regions had via canal systems to Philadelphia.

	Schuylkill Region	Lehigh Region	Wyoming Region
1834	226,692 tons	106,244 tons	70,000 tons
1837	530,152 tons	223,902 tons	115,387 tons
1840	490,596 tons	225,313 tons	148,470 tons [10]

Spread of Mining

The completion of the Upper Grand Section of the Lehigh Navigation between Mauch Chunk and White Haven accelerated the development of anthracite mining. Making the river in the formidable Lehigh Gorge navigable was a feat of engineering that made possible the opening of additional anthracite mines in the Lehigh region. Between 1837 and 1841 the Beaver Meadow,

In 1836 brothers William and Septimus Norris built the George Washington, the first locomotive to self-propel up an incline. Their 4-2-0 anthracite-fueled steam engine climbed the 2,800-ft-long Belmont Incline of the Philadelphia and Columbia Railroad while hauling a tender and a car carrying 24 people. Technological leaps like this soon came to threaten the dominance of canals in transporting heavy goods.

The Norris Locomotive works of Philadelphia was the dominant locomotive producer in America up to the Civil War and was the first to export American locomotives to Europe and South America. It closed in 1866. (NCM/D&L)

Hazleton, Sugar Loaf, and Buck Mountain coal companies began production. All of these mines shipped their coal to market via railroad connections to the Lehigh Navigation Company.

The quantity of anthracite shipped by these newer companies came close to matching, in combination, the tonnage the LC&N moved from its own mines. Of the total 267,826 tons shipped from the Lehigh region during in 1843, the LC&N produced 138,826 tons; the Beaver Meadow Company, 54,729 tons; the Hazleton Coal Company, 44,579 tons; and the Buck Mountain Coal Company, 26,814 tons. An additional 34 tons were produced by small private mines.

The mines of the Beaver Meadow Company, the Hazleton Coal Company, and the Buck Mountain Coal Company were all located in the Eastern Middle Coal Field, and were a considerable distance from the Lehigh Navigation. The fact that all of these companies went to the expense of constructing connecting railroads to the Lehigh Coal and Navigation Company's waterway emphasizes the economic importance of possessing a direct water route during the 1840s to the Philadelphia market, where households and industries had become dependent on anthracite for heat and power.[11]

Alfred Chandler wrote the best summary of the significance of the achievement of the Lehigh Coal and Navigation Company in completing its navigation system and establishing the first commercially successful transportation link between the anthracite coalfields and Philadelphia. Writing in *Anthracite Coal and the Beginnings of the Industrial Revolution in the United States*, Chandler credits Josiah White and Erskine Hazard as the pioneers who created the large-scale anthracite coal-mining industry through their completion of the Lehigh Navigation. He continues:

> The opening of the anthracite fields thus provided the American Northeast with a constantly increasing supply of excellent coal at decreasing prices. Given the existing transportation methods, no other coal source was in a position to provide this massive increase at the same price. The annual output of the Pennsylvania fields rose from 210,000 tons in 1830 to 1,164,000 in 1837, to 1,900,000 in 1844 and to 3,327,000 in 1847. In the two decades between 1830 and 1850, the output of the Virginia fields stayed close to the annual average of 150,000

tons and that of Nova Scotia at 80,000 tons. The Maryland coal fields produced even less than Nova Scotia until the railroads reached them in the late 1840s. Only Great Britain, the world's leading coal producer, did have comparable supplies available. To have met the American consumption of these years, Britain's exports to foreign countries would have had to double and in some years to triple ... Without anthracite coal, the economic developments of the 1830s and 1840s would have been very different. The history of the iron making, iron working, textile, mining, and other industries would not have been the same.

Nor would the economic history of many American states have been the same. Industrial location and industrial output would have been very different and international trade would have followed different times.[12]

All these feats of canal and railroad construction and engineering would have been significant in themselves, for giving birth to civil engineering in the United States and revolutionizing the nation's transportation and commerce. But the anthracite coal they carried ignited an overwhelming change in America: the Industrial Revolution.

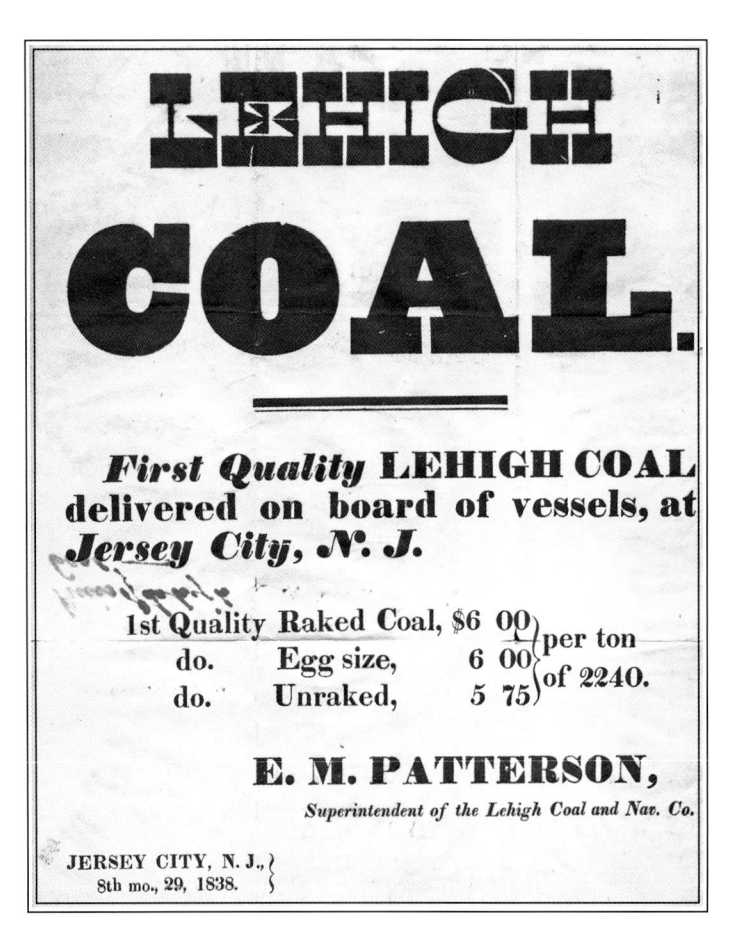

Handbill printed in 1838 announcing the availability of Lehigh anthracite at the port of Jersey City. From here, it would be shipped to ports up and down the eastern seaboard and to inland ports that had canals or riverside docks.

Note the unusual typescript of "Lehigh" and the Quaker-style date in the lower left.

(NCM/D&L)

Chapter Four

Revolutionary Anthracite Iron

"It is of the utmost importance to this Company that the business of making iron with anthracite coal should be established on the Lehigh as speedily as possible ..."

Josiah White

In 1709, an English Quaker named Abraham Darby moved from Bristol, England, up the Severn River to Coalbrookdale, where the remnants of an iron furnace that had been destroyed in an explosion six years earlier still stood. Darby, who had been producing copper pots in Bristol, rebuilt the furnaces and fired them with coke (soft coal from which the impurities have been "baked" out), instead of charcoal, which was becoming both scarce and expensive as hardwood forests in the British Isles disappeared. This made Darby the first ironmaker to use a mineral fuel instead of wood in a commercially successful iron furnace, and effectively ended the use of charcoal in ironmaking in Britain. Moreover, he used his better quality, cheaper iron to mass produce (by early eighteenth-century standards) iron pots and other iron implements with his unique sand-casting process. For these two reasons alone, Abraham Darby the First deserves the title "Father of the Industrial Revolution." Both his son and his grandson, each also named Abraham Darby, made the Coalbrookdale name synonymous with innovations in ironmaking and industrial development in Britain throughout the eighteenth century. Abraham Darby III constructed the world's first cast-iron bridge, the iconic Iron Bridge that still spans the Severn River, in 1778–1780. By the end of the eighteenth century, industrialization's "dark Satanic mills" were spreading rapidly over "England's green and pleasant land."

More than a century after the innovations of the first Darby, the United States was still an almost completely agrarian society. Such manufacturing as there was in America was small, slow, and wood-based, both in its machinery and its fuel. "The lack of metal and metal machinery, as well as the small production of iron, reflected the backward state of the technology of the American iron industry," wrote A.D. Chandler in 1972. "In the United States, iron manufacturers were still producing small amounts by ancient techniques."[1]

It was not as though Americans were ignorant of the industrial revolution that had taken place in England. Leaders in the scientific, business, and political communities were well aware of the economic and industrial potential of successfully smelting iron with mineral fuel. Beginning in 1825, the Franklin Institute offered a gold medal to anyone who produced at least twenty tons of iron smelted using mineral fuel. Nicholas Biddle, president of the Bank of the United States, and some of his business associates offered a $5,000 prize to the first person keeping a furnace in blast using anthracite alone for at least ninety days. White and Hazard themselves, realizing the potential bonanza for their anthracite business if the coal could be used for industrial purposes as well as in blacksmiths' forges and home heating, experimented unsuccessfully with using anthracite to smelt iron at a furnace in Mauch Chunk in 1826.

In 1835, the state provided for a geological survey, primarily to learn more about Pennsylvania's coal and iron deposits. Also in 1835, *Hazard's Register of Pennsylvania* suggested that the Pennsylvania legislature award premiums for producing pig iron with anthracite coal. On June 16, 1836, a bill sponsored by Charles B. Penrose of Cumberland, Perry County, was enacted to encourage the manufacture of iron with coke or mineral coal by authorizing the governor to form a joint stock company for that purpose. But even with increasing numbers of skilled ironworkers and mechanics from England, Scotland, and Wales emigrating to the United States, bringing with them their experience and knowledge of the most modern British practices, finding a highly productive way to use hot-burning anthracite in an iron furnace remained elusive.[2]

White and Hazard Re-Enter the Contest

In February 1837 George Crane, owner of the Yniscedwin Iron Works in Wales, and his furnace superintendent, David Thomas, for the first time combined the hot-air-blast process, patented by Scotsman James Neilson in 1829, with anthracite from a seam adjacent to the works. Thomas's twenty years of diligent, though sporadic, experimentation with anthracite was finally successful, and the innovative combination produced good quality iron at the rate of 34 to 36 tons per week.

Crane patented "the application of anthracite or stone coal and culm, combined with the using of hot air blast in the smelting and manufacture of iron." While his patent was published in the *Journal of the Franklin Institute* in 1838, thereby making his invention public knowledge among American ironmakers, it is certain that Josiah White and Erskine Hazard of the Lehigh Coal & Navigation Company had already received the news. White's nephew, Solomon W. Roberts, had toured Welsh ironworks in 1837, inspecting rails for the newly formed Philadelphia & Reading Railroad. Roberts visited Mr. Crane's establishment in May 1837 for the purpose of seeing the new process and satisfying himself that the materials used were similar to those that exist so abundantly in Pennsylvania. Roberts later wrote: "Finding that the great object was accomplished, and that the results were highly gratifying, I communicated the fact to his [uncle Josiah White's] friends in Philadelphia, by whom it was shortly after made public through the newspapers."[3]

The news from Wales stimulated renewed efforts by White and Hazard to smelt iron using anthracite. The experimental furnace constructed in 1826 was reactivated, the new experiments being conducted by the firm of Baughman, Guiteau & Company. In a letter to Walter R. Johnson dated November 9, 1840, F.C. Lowthrop, who later became a member of that firm, recalled:

View of the Ynescedwyn Iron Works in Wales, 1838. The owner, George Crane, and his ironmaster David Thomas combined anthracite from nearby mines with the hot-blast technique pioneered by Scotsman James Neilsen in an enlarged furnace, and by early 1837 were turning out 34 to 37 tons of good quality iron per week. *(NCM/D&L)*

During the fall and winter of the year 1837, Messrs. Joseph Baughman, Julius Guiteau and Henry High, of Reading, made their first experiment in smelting iron with anthracite coal, in an old furnace at Mauch Chunk, temporarily fitted up for the purpose; they used about eighty per cent of anthracite, and the result was such as to surprise those who witnessed it (for it was considered as an impossibility even by ironmasters); and to encourage the persons engaged in it, to go on. In order, therefore, to test the matter more thoroughly, they built a furnace on a small scale, near the Mauch Chunk Weigh Lock, which was completed during the month of July 1838.

This experimental furnace was only 21½ feet high by 22 feet square at the base; the bosh was 5½ feet across. The two blowing cylinders were wood, each 6 feet in diameter; they were powered by a 14-foot overshot water wheel. Temperature of the blast did not exceed 200°F. During the autumn of 1838, an improved heating apparatus was procured. The blast was again applied about the end of November 1838, and the furnace worked remarkably well for about five weeks, fueled exclusively with anthracite coal. The average blast temperature was about 400° F. The following season additional modifications were made. The furnace was blown in again on July 26, 1839, and continued in operation until November 2 of that year, worked exclusively with anthracite. The temperature of the blast varied between 400° and 600° F.[4]

While Baughman, Guiteau & Company were the first to achieve any real success with anthracite in this country, their results were not wholly satisfactory. The furnace was strictly experimental and was incapable of producing commercial amounts of quality iron for long periods.[5] The managers of the Lehigh Coal and Navigation Company concluded that the best course of action was to tap the expertise of Crane and Thomas, so they sent Erskine Hazard to Wales to negotiate with George Crane. Hazard, accompanied by his eldest son, Alexander, arrived in Wales in November 1838. By the time of Hazard's arrival, Crane was the holder not only of the British patent for making anthracite iron, but also of the U.S. patent, which he had acquired after the death of its holder, Dr. Frederick Geissenheimer, earlier that year.

It is extremely doubtful that White and Hazard ever expected Crane to leave his interests in Wales. His furnaces were profitable and he expected his patents to have similar success. Instead, it is more reasonable to assume that Hazard planned to discuss royalties and also to engage someone who could duplicate Crane's achievement in Pennsylvania. In a remarkably unselfish move, Crane suggested his superintendent, David Thomas. Thomas certainly had the experience and qualifications for the task as he had spent all his adult life in the iron trade and had supervised the modifications to the cupola furnace in which anthracite iron had first been made.

David Thomas, Ironmaster

Apparently satisfied with Thomas's credentials and personality, Erskine Hazard offered him the job of building an anthracite iron furnace on the Lehigh Navigation. Thomas was at first reluctant to leave his native land, chiefly on account of his aged mother, but his wife, Elizabeth, persuaded him that the United States held larger opportunities for their three sons.

On December 31, 1838, Thomas and Erskine Hazard signed a contract to bring Thomas to America to work for the new Lehigh Crane Iron Company. It consisted of four clauses. First, Thomas must agree

> ... to remove with his family to the works to be established by the said company on or near the river Lehigh, and there to undertake the erection of a blast-furnace as furnace-manager; also to give his assistance in finding mines of iron-ore, fire-clay, and other materials suitable for carrying on iron-works, and generally give his best knowledge and services to the said company, in the prosecution of the iron-business, in such manner as will best promote their interests, for the term of five years from the time of his arrival in America, provided the experiment of smelting iron with anthracite coal should be successful there.

Secondly, the company agreed to pay the moving expenses of Thomas and his family

> ... from his present residence to the works ... on the Lehigh, and there to furnish him a salary at the rate of two hundred pounds sterling a year from the time of his stipend ceasing in his present employment until the first furnace on the Lehigh is got into blast with anthracite coal and making good iron, and, after that, at the rate of two hundred and fifty pounds sterling shall be added to his annual salary.

Thomas would receive an additional fifty pounds a year for each additional furnace put into blast under his management. Within the first ten years of his employment, he built five furnaces for the Lehigh Crane Iron Company.

The third clause provided for failure. It was mutually agreed

> ... that should the said Thomas fail of putting a furnace into successful operation with anthracite coal ... the said company shall then pay the said Thomas a sum equivalent to the expense of removing himself and his family from the Lehigh to their present residence.

The fourth clause set the exchange rate for settling Thomas's salary at four shillings and sixpence to the dollar.* The agreement was signed by David Thomas and by Erskine Hazard "for Lehigh Crane Iron Comp'y" and witnessed by Alexander Hazard, on December 31, 1838.[6]

Prior to Hazard's journey to Wales, the Lehigh Crane Iron Company existed only in the minds of its promoters, principally White and Hazard. On January 10, 1839, the first meeting of its directors was held in Philadelphia, at which time Robert Earp was elected president and treasurer and John McAllister Jr., secretary. The other directors were Josiah White, Erskine Hazard, George and Thomas Earp, and Nathan Trotter.

On October 2, 1838, in anticipation of the organization of the Lehigh Crane Iron Company, and to promote the development of other iron-producing and coal-consuming enterprises along their canal, the board of managers of the Lehigh Coal & Navigation Company adopted a resolution to provide free water rights to those building anthracite-fueled iron works. The offer was limited to the region between Allentown and Parryville because it was the most undeveloped section of the navigation. This offer was a significant concession because the LC&N was reaping considerable revenue from the sale of waterpower rights downstream from Allentown, especially in Easton's Abbott Street area, where waterpower from the canal had attracted a number of industries including a saw mill, a grist mill, a cotton mill, a nail works and a gun factory.

At their meeting on January 3, 1839, the company's managers acknowledged receipt of a letter dated November 9, 1838, from a group consisting of Erskine Hazard, Josiah White, Robert, George and Thomas Earp, Thomas Mitchell, John McAllister Jr., and Jonathan K. Hassinger, of whom all but the last were to become directors of the Lehigh Crane Iron Company, accepting the offer contained in the October 2 resolution. However, as neither this group, nor any other, had met the company's conditions by the deadline of July 1, 1839, and recognizing that "it is of the utmost importance to this Company that the business of making iron with anthracite coal should be established on the Lehigh as speedily as possible," the managers of the Lehigh Coal and Navigation Company met on July 2, 1839, to rescind the first resolution and pass a second in revised form.[7]

The revised resolution made no reference to George Crane of Wales; it simply specified that the iron had to be smelted with anthracite coal. The initial capital requirement was reduced from $50,000 to $30,000. While the production rate of 27 tons per week was retained, the three-month requirement was eliminated. The other revisions were minor. The deadline on this new resolution was September 1, 1839. On July 8, 1839, the directors of the Lehigh Crane Iron Company accepted the company's offer. On December 15, 1840, the Lehigh Crane Iron Company, having

---

* The 1838 value of one pound was approximately $113 in 2018 values. One British pound contained twenty shillings; each shilling contained twelve pence.

met all the imposed conditions, was granted a conveyance of the property and water rights by the Lehigh Coal and Navigation Company.[8]

Thomas Arrives in America

Thomas spent the first months of 1839 ordering the machinery he would need in Pennsylvania from shops and foundries in Wales, because the necessary equipment could not be had in the United States. He and his family left Wales early in May, and crossed the Atlantic in 23 days, a near-record time. After a month-long stay on Staten Island, the family set out for Allentown on July 6 and arrived, via the New Jersey railroad and various stage coaches, three days later. On July 11, David Thomas and his oldest son, Samuel, walked to Biery's Port, the site the company had purchased for the new furnace.

Biery's Port was a small group of houses owned by various members of the Biery family, clustered around a 1762 grist mill. Nearby stood a Georgian-style house built in 1768, and briefly occupied by George Taylor, a signer of the Declaration of Independence who had been the ironmaster at Durham Furnace in Bucks County. The tiny hamlet was surrounded by fields farmed by families who, like the Bierys, spoke mostly German. The presence of the grist mill had made Biery's Port (or Biery's Bridge, as it was also known) something of a transportation hub; a toll bridge crossed the Lehigh from Whitehall Township there, and linked the west bank of the Lehigh with a road that went east to Bethlehem.

David Thomas

This was where Thomas was to build his furnace. The Lehigh Coal & Navigation Company had purchased a tract of land adjacent to Lock 36 from Solomon Biery. Lock 36 was the first lock below Dam Number 6, and the fall of the level between the two was eight feet. This meant that enough current could be generated to power water wheels to drive Thomas's blowing engines for the hot blast. David and Samuel Thomas took measurements on July 11, 1839, in what was still a farm field.

Actual construction on the first furnace for the Lehigh Crane Iron Company commenced about August 1, 1839. The stack was 45 feet high from the base of the hearth to the trunnel head, with a chimney extending about 12 feet above the trunnel head. It was constructed of limestone, 30 feet square at the base and tapering to about 23 feet square at the top. It was lined with two courses of 9-inch-thick firebrick with clay packing 2 to 3 inches thick between the lining and the stone, to provide for expansion of the lining. The hot-blast stoves, with the usual bed pipes, consisted of 4 ovens of 12 arched pipes each, 5 inches interior diameter, 1½ inches thick in the legs and 2 inches thick in the arch. They were built on the ground and fired with anthracite.

David Thomas, 1794–1882

David Thomas was born in Glamorganshire, South Wales, and after leaving school at age 17, was apprenticed at the Neath Abbey Iron Works. In 1817, at the age of 22, he became superintendent of the Yniscedwyn Iron Works. During a few years of sporadic ironmaking due to mismanagement by the works' owners, Thomas surveyed and built the Brecon Forest tramway and helped open the Drim anthracite mine. When the Yniscedwyn works were bought by hardware magnate George Crane in 1823, Thomas returned full time and, encouraged by Crane, began experimenting with smelting iron with anthracite coal. As had happened at the same time in Pennsylvania, Thomas's efforts were unsuccessful until 1837, when he and Crane perfected the hot blast method that had been developed by James Neilson. Crane patented their method, and also bought the rights to a similar process that had been developed by American Frederick Geissenhainer, who had died before being able to perfect his own patented hot-blast process. The process allowed high-quality iron to be produced at the rate of 34 to 36 tons per week, and caught the notice of the managers of the Lehigh Coal and Navigation Company.

Thomas's success with the Lehigh Crane Iron Company, builders of the first technically and commercially successful anthracite iron furnaces, earned him the title "Father of the American Anthracite Iron Industry." In Catasauqua, the town he founded around his iron works, he was known simply as "Father Thomas." Coming from the British tradition of paternalistic manufacturing communities, Thomas built a church and a public waterworks while he was constructing his first furnace. Company-owned brick houses with slate roofs soon lined the first streets, and a community bakeoven and a fire department were established.

For most of the 1840s, the town, known then as Craneville, was an island of mostly single Welsh-speaking men in a sea of German-speaking farmers. But as the iron business took off, new industries and people came to town: churches, schools, stores, and more company houses filled the former farm fields. In 1853, when the town was incorporated as a borough and renamed Catasauqua, David Thomas was elected the first burgess (an old title for a mayor) and authored many of the first ordinances for the town.

In 1855, David Thomas resigned as superintendent of the Lehigh Crane works in order to help develop the Thomas Iron Company's first works, across the Lehigh River. Again, a company town was built; in this case Thomas, an ardent opponent of alcohol, succeeded in keeping it dry. Like Catasauqua, Hokendauqua got its Presbyterian church, its school and stores, but without an excellent water system to attract brewers, all its places for workers to wet their whistles after work were on the outskirts of town.

Thomas was among the founders and was the first president of both the American Association of Industrial Engineers and the American Metallurgy Society. Greatly honored in his lifetime, his funeral in 1882 is said to have been the largest event ever held in Catasauqua. The head of his mile-long funeral procession, comprised of prominent men of industry and government as well as virtually every resident of Catasauqua, reached Fairview Cemetery across the Lehigh before those in the end even started to walk. Despite this acclaim, today his only memorial is the unobtrusive, grass-covered mound that is the Thomas family crypt.

Furnace No. 1, Lehigh Crane Iron Works, 1840

No contemporary images or diagrams remain of the Lehigh Crane Iron Works' Furnace No. 1, the first commercially and technically successful anthracite-fired hot-blast iron furnace in North America. This artist's rendering by Dennis Gerhart is drawn from descriptions and diagrams included in a memoir by Thomas's son Samuel in 1899. Mr. Gerhart was assisted in his research by Craig L. Bartholomew and Donald S. Young.

Built at Lock 36 on the Lehigh Navigation, the furnace was constructed of limestone and was 45 feet high, with a 12-foot chimney. It was 30 feet square at the base, tapering to 23 feet square at the top. It was lined with two courses of 9-inch-thick firebrick with 2–3 inches of clay between the lining and the stone to allow the firebrick to expand. A water-powered elevator brought the coal, iron ore, and limestone to the top of the furnace.

The blast was supplied by a water wheel 12 feet in diameter and 24 feet long, driven by water from the Lehigh Canal that was diverted into a channel that dropped eight feet. The wheel turned gears, which pushed beams on a frame that drove the blowing cylinders. The cylinders, five feet in diameter with a six-foot stroke, forced air through ovens in small buildings on the ground alongside the furnace. The air, heated to 600°F, was then blown into the furnace, where the burning anthracite raised the temperature to 2,700°F or higher. Molten iron was tapped from the bottom of the furnace and guided to flow into channels in sand on the casting floor.[9]

(*Dennis Gerhart for NCM/D&L*)

In this cut-away view by Dennis Gerhart of a mid-nineteenth-century anthracite iron furnace, anthracite coal, iron ore, and limestone are "charged" (dumped through the "trunnel head" at the top of the furnace) into the blast furnace in that order. Superheated air is blasted into the lower part of the furnace (the crucible) through openings called "tuyeres."

As the burning anthracite and the air from the hot blast mix with and melt the ore and limestone, the minerals descend through the furnace, heated to 2,700°F. The heavier molten iron dropped to the "hearth" at the bottom, where it was drawn off by workers who drilled through or knocked out a clay plug, allowing the molten iron to flow into troughs of sand on the casting floor. This was the "casting bed" and was formed with wet sand that had been left to dry. The shapes moulded in the troughs were iron bars called "pigs." The lighter "slag," created when the limestone bound up impurities in the iron ore, floated on top of the iron and was drawn off and dumped away from the iron works.

(Dennis Gerhart for NCM/D&L)

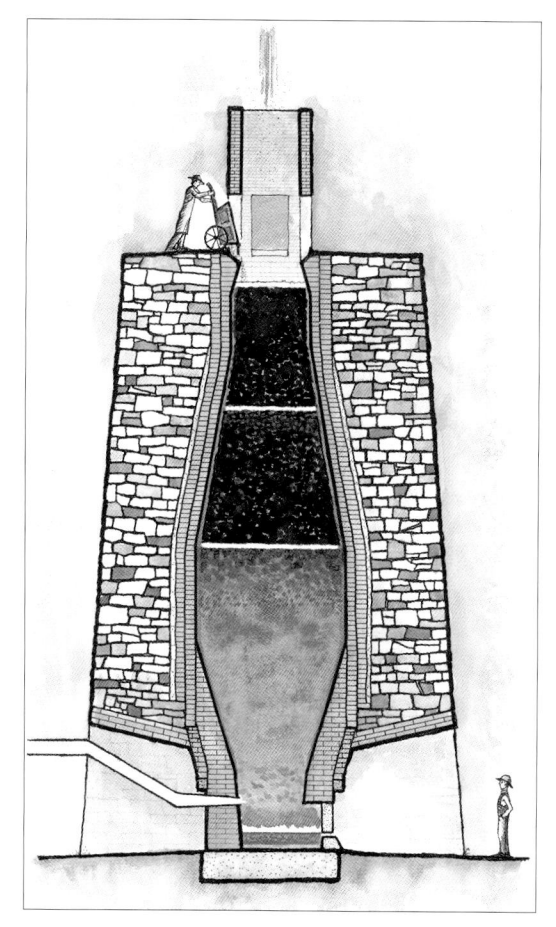

Although he had built furnaces in Wales, starting with such an empty canvas in Pennsylvania must have been challenging for Thomas. For one thing, he was presumably accustomed to having his blowing engines powered by steam engines, which had long since replaced water wheels in Britain. The challenges mounted when the equipment ordered in Wales arrived, via the Delaware and Lehigh canals from Philadelphia, but minus the blowing cylinders: these were too large to load onto a ship. With no foundry in the nation capable of boring a cylinder 60 inches in diameter, Thomas had to persuade the Southwark foundry in Philadelphia to enlarge their boring mill to make the cylinders he needed.[10]

Eleven months later, Thomas triumphed. He later wrote:

> After many vexatious delays, the furnace was completed and successfully blown in at 5 o'clock P.M. July 3, 1840, and the first cast of about 4 tons of iron was made on the memorable 4th of July of that year, the keepers in charge of the furnace being William Phillips and Evan Jones.[11]

The proportion of ores used was about one-fourth magnetic and three-fourths brown hematite (limonite). The magnetite came by canal boats from the Irondale, Byram, and Dickerson mines in New Jersey and the hematite was hauled in wagons from Rice's mine, near Schoenersville, Hanover Township.

By the mid-1850s, the Crane Iron Company had five furnaces at Catasauqua and David Thomas and son Samuel were preparing to establish the Thomas Iron Company at Hokendauqua. Son John remained at Crane as superintendent. Thomas Iron was to become the largest merchant pig producer in the Lehigh Valley, with furnaces in several locations.

(Courtesy of Presbyterian Church of Catasauqua)

The furnace remained in blast until its fires were quenched by the rising waters of the great flood of January [7 & 8], 1841, a period of six months, during which 1080 tons of pig-iron were produced. The largest output for one week was 52 tons.[12]

The furnace built by David Thomas at Biery's Bridge (now Catasauqua) was not the first to make iron with anthracite coal in the United States, but it was the first to be both a technological and a commercial success. For the first time in this country, large quantities of high-quality iron could be made quickly. William Firmstone, builder and longtime superintendent of the second anthracite iron furnace, the Glendon Iron Works, asserted 34 years later in his "Sketch of Early Anthracite Furnaces,"

> With the erection of this furnace commenced the era of high and larger furnaces and better blast machinery, with consequent improvements in yield and quality of iron produced.[13]

More than any other single achievement to that time, it signaled the beginning of the industrial revolution in America.

Chapter Five

1840 TO 1850
THE IRON AGE

"Catasauqua, this vigorous little Vulcan of the valley..."

Mathews and Hungerford, 1884 History of Lehigh and Carbon Counties

The years following David Thomas's success at Catasauqua in 1840 witnessed the rapid transformation of the region—indeed, of the entire Northeast—from a comparatively tranquil agrarian society to a more tumultuous industrial one. Locomotives began to replace the slower canal boats as the principal means of transportation. Steam engines instead of water wheels powered the new mills and factories. While some more conservative observers deplored these new developments, others saw in them progress, romance, and adventure.

Lehigh Valley businessmen and investors were quick to appreciate the potential profit in anthracite ironmaking. Within a few years after David Thomas's commercial success at Catasauqua, numerous other companies were formed and furnaces were soon under construction all along the Lehigh and some of its tributaries. During the 1840s, anthracite-fueled furnaces were constructed along the Lehigh Navigation at Glendon (1844), South Easton (1845), Allentown (1846), and along the Delaware Division Canal in Bucks County at Durham in 1848.

At the peak of anthracite-iron production, 55 blast furnaces were in operation at 21 locations in the greater Lehigh Valley. Both transportation economies and the lower price of coal contributed to the lower price of anthracite iron. Eastern Pennsylvania—particularly the Lehigh and Schuylkill valleys—became the center of iron production in the United States.

It would be difficult to overemphasize the importance of anthracite ironmaking in the continuing development of the American iron and steel industry. While it was not the ultimate technology, the new practices and machinery developed by the ironmakers of this era established the foundations on which all later improvements were based. In 1810, total United States iron production was 53,908 tons, of which 26,870 tons were produced in Pennsylvania. The average output per American furnace was only 352 tons a year. By 1840, there were 804 iron furnaces in the U.S., of which 213 were in Pennsylvania. Total U.S. production in 1840 was 286,903 tons, of which 98,395 were made in Pennsylvania. While total production had increased fivefold, the average production per furnace had risen only to 357 tons. After 1840, production increased dramatically. By 1847, the 304 furnaces in Pennsylvania produced 389,350 tons of pig iron, more than was produced in the whole country only seven years earlier. The use of anthracite was largely responsible for this tremendous increase.[1]

By the time of the Civil War, the anthracite-fueled furnaces of eastern Pennsylvania were producing half of the nation's iron; the Lehigh Valley alone produced slightly more than 20 percent of the national total.[2]

Glendon Iron Furnace

The Glendon Iron Company was the second anthracite-iron works established in the Lehigh Valley. In 1842, Boston entrepreneur Charles Jackson Jr. hired a noted English ironmaster named William Firmstone to assess the potential for ironmaking in the Lehigh Valley. After a visit that included a stop at the Crane Iron Works, Firmstone reported that not only was the area the best suited for iron production in the entire state, but the Lehigh Coal and Navigation Company was eager to support industrial development along its canal. Authorized by Jackson, Firmstone bought a site two miles west of Easton along Section 8 of the Lehigh Canal, located between the canal and the south bank of the Lehigh River. (This site is now Easton's Hugh Moore Park and Section 8 is the only fully restored section of the Lehigh Navigation.)

William Firmstone, 1810–1877

Firmstone learned the iron business in his uncles' furnace in Dudley, England. He emigrated to the United States in 1835; after working at furnaces in Ohio and Pennsylvania, he was hired by a group of Boston investors to build an anthracite-iron furnace. As David Thomas recommended, he purchased land between the Lehigh River and canal near Easton from the Lehigh Coal and Navigation company in 1842. Recognized as both an innovative ironmaster and an astute manager, Firmstone oversaw the iron works until his death in 1877. He is buried in Easton Cemetery.

Construction began in the fall of 1842, and by 1844 the first production furnace was blown in. As with the Crane Iron Company, the blowing engines were powered by water wheels driven by the canal; like Crane, the Glendon Works were located about a mile downstream of one of the Navigation Company's dams, and the current in the canal could be tapped to drive water wheels.

Furnace No. 2 was constructed and blown in the next year; in 1846, the two Glendon furnaces produced over 7,000 tons of iron. No. 3 furnace was constructed in 1850.

An early photograph of the Glendon Iron Company, with Section 8 of the Lehigh Canal in the foreground. The Lehigh River is on the other side of the structures. The Glendon furnaces received magnetic ore from New Jersey by canal boat. This was the one of the largest and most important industries in the industrial park developed where Hugh Moore Park is today.
(NCM/D&L)

Firmstone was as expert an ironmaster as David Thomas, though he came much later to the anthracite-fired, hot-blast method than Thomas did. Under his direction, the Glendon Works established an excellent reputation. Like most other Lehigh Valley iron furnaces, Glendon used both New Jersey magnetite ore and locally sourced brown hematite. The magnetite was brought in via the Morris and Lehigh canals, and the brown hematite came from mines in bordering Williams Township.

Most of Glendon's pig-iron production was shipped via canals to New York, where it was loaded onto coastal schooners for shipment to Boston. There it was rolled into bars and rolls at the Glendon Rolling Mill near Boston and sold throughout New England.[3] This broke the New England metal-working industries' dependence on iron imported from Britain, and greatly expanded the demand for domestically produced iron.

As did Thomas in Catasauqua, Firmstone drew on his heritage of British industrial paternalism by building housing for his workers. The tiny borough of Glendon, built along the hillside south of the iron works and the Lehigh Canal, survives 120 years after the demise of the company that founded it.

Durham Iron Furnace

The first Durham Furnace was built in 1727 on Durham Creek about a mile west of where the creek empties into the Delaware River. It was in full operation by the next year. The pig iron produced there was shipped to Philadelphia in vessels that became known as Durham boats.

Durham Furnace, like every other iron furnace in the American colonies, used charcoal and a cold-air blast. Its early owners included some of the most prominent people in colonial Pennsylvania, including Anthony Morris, William Allen, Joseph Turner, James Logan, Israel Pemberton, and Joseph Galloway. Durham Furnace produced shot and shells for both British and Colonial forces fighting in the French and Indian War and for Washington's troops during the American Revolution. During both conflicts, Durham was managed by George Taylor, who was Justice of the Peace (in effect, the county executive) of Northampton County, which at that time included what is now Lehigh, Carbon, and Monroe counties, and was a signer of the Declaration of Independence.

The original furnace operated until 1789. By 1791, the site was largely abandoned and much of the equipment sold to the Hibernia Furnace in New Jersey. Early in the nineteenth century, a

This circa 1855 drawing of the second Durham Furnace shows the Bucks County iron works a few years after it was rebuilt as an anthracite iron furnace along Durham (Cooks) Creek, near the Delaware Canal. (NCM/D&L)

grist mill was built on the site of the original furnace.

Joseph Whitaker and Company built two anthracite-fired furnaces near the mouth of Durham Creek in 1848-1849. Each was capable of producing 5,000 tons of iron a year. The anthracite fuel for the furnaces was brought on the Delaware Canal, and the pig iron produced was carried away by the same route.

Control of the Durham Furnace remained with the Whitaker family until 1864, when it was sold to Edwin Cooper and Abram Hewitt for the first time.* Cooper and Hewitt, the first a brilliant engineer and ironmaster, and the second a visionary businessman, brought major innovations to the entire iron industry starting in the 1870s, and will be further discussed in Chapter Eight.[4]

Water, Coal, and Iron

The three iron works described above, as well as the Allentown Iron Company, which was established in 1846 on the west bank of the Lehigh River just above the site of the present Tilghman Street bridge, all initially relied on the canals to get their fuel from the coal regions and much of their iron ore from New Jersey, to get waterpower from the canals to drive their blowing engines, and for canal boats to carry their pig iron to market. While they were producing large quantities of high-quality iron by what were then the most modern techniques, they still depended on the ancient power of water as well. The coming of the railroads and the use of stationary steam engines in place of water wheels made it possible for iron works to be established away from water routes. This triggered a massive expansion of the industry in the Lehigh Valley and beyond in the late 1850s.

Changing the Face of the Lehigh Valley

The ironmaking boom had a profound effect on the character and lifestyle of the Lehigh Valley and its inhabitants. New towns and villages were established, and a number of those already in existence began to grow rapidly. Some communities—Catasauqua, Coplay, Hokendauqua, Glendon, and Alburtis—owe their existence to the iron industry. Others, like Emmaus, Macungie, and Hellertown, founded decades earlier, grew rapidly in size and prosperity after furnaces were

---

* They sold it the following year, and bought it back at a sheriff's sale in 1870.

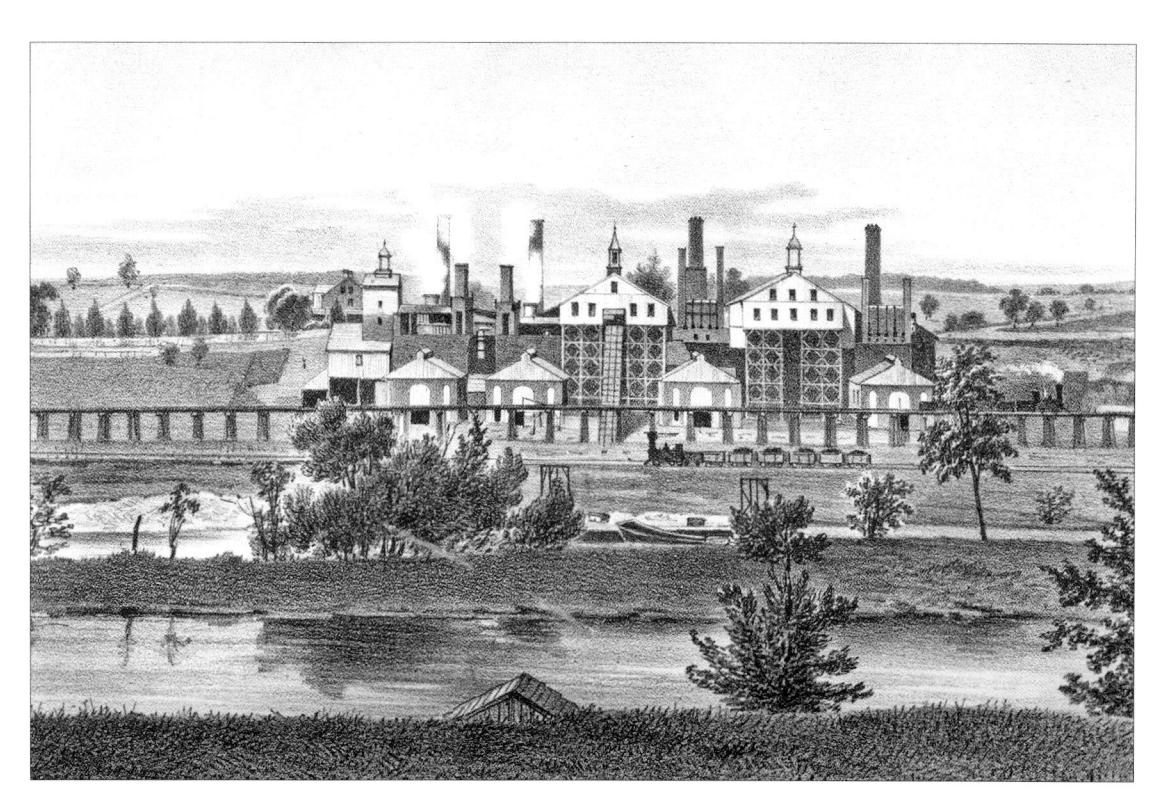

View published in 1860 of the Allentown Iron Works, a large complex along the Lehigh River that had four furnace stacks by the time of this image. Early locomotives can be seen. One is on the trestle; the other appears to be traveling south pulling five cars filled with anthracite. The only railroad constructed by this time was the Lehigh Valley Railroad, the chief competitor of the Lehigh Coal and Navigation Company for transporting anthracite. Today, the American Parkway bridge passes over the site. (M.S. Henry, History of the Lehigh Valley)

built in the town or nearby. Across the river from Bethlehem, what had been Moravian community-owned farms in 1845 had become, by 1860, a community of 1,000 people clustered around a zinc works and two railroads in what was to become South Bethlehem and, subsequently, the site of the Bethlehem Iron Company. Allentown, too, changed and grew. In addition to the furnaces and rolling mills of its two iron companies, a host of other industries were established there to provide parts, equipment, and mill supplies to the furnaces, and to use their pig iron.

The anthracite-iron industry triggered major changes in the ethnic character of the Lehigh Valley. Before 1840, English was rarely spoken in the countryside, and Pennsylvania Germans still dominated the economic, cultural, and social life of the communities. The rapid growth of the iron industry and the related expansion of the railroad and coal-mining industries created more jobs than there were workers to fill them. A new wave of immigrants arrived, mostly from England, Scotland, Wales, and Ireland. Most, especially the English, Scots, and Welsh, already had the training and experience needed for the more technical and skilled jobs in the furnaces. The Irish, at least for a while after they arrived, were relegated to the dirty and dangerous work around the furnaces, and to building railroads.

The new immigrants at first tended to adopt local culture and customs. Writing to a friend in Wales in December 1839, David Thomas relayed the following impressions of his new home:

> We live in a fertile country where every sort of grain, vegetable and fruit is abundantly grown. The climate is very healthy; and the weather has been hitherto very good. The people are hospitable and kind, chiefly from German origin. There is much of that language spoken here, which I am learning very fast. The children can talk it better than I can.[5]

As the years passed and they became more numerous, these immigrants from the British Isles increasingly began to influence the local culture. Their impact on rural Pennsylvania-German farmers was minimal, but in the larger towns and cities it became considerable. It was enhanced by the fact that many of them, and their countrymen, had become wealthy in the iron and related industries. Because they were English-speaking, they had far less difficulty assimilating into the prevailing culture than immigrants who arrived in the later decades of the nineteenth century.

Iron Ore and Limestone Mining

The Lehigh Valley had abundant reserves of iron ore and other resources. While some mines were opened in the Lehigh Valley region during the eighteenth century to provide ore to nearby charcoal furnaces, it was not until the development of anthracite technology and the resulting demand for vastly larger quantities of raw materials that this natural resource was exploited in

The Thomas Iron Company acquired mines containing ore that had the chemical composition it wanted in its furnaces. These photos are both of the Koch brown hematite mine near Hellertown. The view on the right shows the track that carried ore up to the washery, where clay and mud were washed off. Clay was always found in limonite mines mixed in with the ore. Wash water was very often disposed of in areas where the ore had been extracted; sometimes is was simply directed onto fields where it formed huge puddles. On the left is a detail of the same mine, showing men and mules at work loading ore buggies. These photos are among many in the 1904 report from the company to its stockholders.

This washery and other ground-level workings were at a mine that provided ore to the Macungie Furnace, a small furnace in East Macungie. For several decades sights such as this were common throughout the limestone areas of the Corridor, where pits were opened on numerous farms in search of the valuable ore.

(Courtesy of Lower Macungie Township Historical Society)

earnest. Brown hematite ores, called limonite in modern terminology, were mined early in the nineteenth century and carted to the few charcoal furnaces scattered throughout the region. During the explosive growth of the anthracite-iron industry after the year 1840, the search for ore was to change the landscape of Lehigh County in particular.

During the 1840s, numerous mines were opened in the Lehigh Valley. Ore was hauled in wagons over dirt tracks to the new furnaces along the Lehigh River. By the 1850s, as many people were employed mining ore as were tending the furnaces.

At various locations between Emmaus and Vera Cruz in Lehigh County, also at Siesholtzville and Rittenhouse Gap in Berks County, mines opened in the early anthracite era were operated until the early years of the twentieth century. The limonite or brown hematite ores that were used so extensively during the early years of the anthracite-iron industry were found chiefly above the extensive limestone formations found throughout the full length of the Valley proper and also in the Saucon Valley. Many very productive mines were also developed near the base of the northern slope of the South Mountain in Lehigh County, and in the high valleys of Williams Township and eastern Lower Saucon Township in Northampton County. The first, associated with the limestone, were referred to locally as "valley ores" and those that occurred at the higher elevations were called "mountain ores." Chemically, they were essentially the same.

Most of the limonite ores lay relatively close to the surface and were mined principally by the open-cut or open-pit method. This method proved quite satisfactory during the early stages of operation but as some of the richer deposits were mined to considerable depths, the loose overburden tended to slide into the pits after heavy rains. Therefore, mine operators sometimes resorted to shaft mining, sinking shafts in or near the old pits. Drifts were also cut into the slopes of the mines to follow the ore.

The pattern of ownership of the limonite deposits was diverse. All of the local iron companies owned and operated at least three or four mines. Others were owned by small local mining companies and farmers who operated one or more mines on their own property. The large multi-furnace companies, such as the Thomas Iron Company, Crane Iron Company, Allentown Iron Company, Bethlehem Iron Company, and Glendon Iron Company, often owned and operated dozens of mines. By doing so, they had direct knowledge of operating costs. Because they were also the major customers of the independent operators, the large iron producers effectively determined the market price of ore and the wages of the miners. In times of strong demand for iron, ore prices, wages, and profits were high. When times turned "dull" prices fell, wages were cut, and profits disappeared. The less productive mines were then shut down, sometimes forever, but often to return to operation when iron prices rebounded.

Magnetite-ore mining was never as extensive in the Delaware and Lehigh National Heritage Corridor as was the mining of the brown hematites. The Thomas Iron Company's mines at Rittenhouse Gap and the Reading Coal and Iron Company's mines near Seisholtzville were fairly large operations but were never as productive as the mines in the New Jersey section of the Reading Prong. All of the local furnaces used New Jersey magnetic ores, transported first by the Morris Canal and later by the Central Railroad of New Jersey to Phillipsburg, New Jersey. From there the ore was transported to the Lehigh Valley furnaces—originally on the Lehigh Navigation and later by the Lehigh Valley, the East and North Penn branches of the Philadelphia and Reading Railroad, or by the Central Railroad of New Jersey itself.[6]

Coal from the Wyoming Valley

The establishment and rapid growth of the anthracite-iron industry greatly boosted the demand for coal. The completion of the Lehigh Coal and Navigation Company's transportation link between the Wyoming Valley and the Lehigh River by 1844 opened up the even greater coal resources in the Wyoming Valley. In addition, the fact that Wyoming Valley coal lay largely in horizontal beds, which made it easier to mine, brought about a surge in the Wyoming region's anthracite production between 1844 and 1852. The relative production of the three competing regions for the years 1844, 1848, and 1852 makes clear the growth in Wyoming Valley production:

	Schuylkill Region	Lehigh Region	Wyoming Region
1844	887,937 tons	377,002 tons	365,911 tons
1848	1,733,721 tons	670,321 tons	85,196 tons
1852	2,636,835 tons	1,072,136 tons	1,284,500 tons [7]

The Wyoming region equaled and then surpassed the Lehigh region as an anthracite producer during the late 1840s, as the above figures document. Much of this increased production was transported via the Lehigh Coal and Navigation Company's transportation system in order to reach Philadelphia. After 1844, the Lehigh Coal and Navigation Company's transportation system became the first and only commercial route to Philadelphia used by producers in the northern middle and lower coalfields.[9]

East Sugar Loaf Mines, opened in 1850 along the Hazleton Railroad at Stockton, two miles from Hazleton, were in the northern coalfield in Luzerne County. Part of the company village is seen in this 1860 engraving. Mining took place too close to the surface, and in 1869 a mine caved in, swallowing some houses. Ten people died in the disaster, as did an unreported number of miners.[8] Building and road collapses and cave-ins resulting from mine subsidence and poor mining techniques persisted well into the twentieth century in the anthracite regions.

(M.S. Henry, History of the Lehigh Valley)

Wire Rope

Like so many other things in the industrial history of the Delaware and Lehigh National Heritage Corridor, the origins of the wire rope industry can be traced to the activities of Josiah White and Erskine Hazard.

During the winter of 1815–1816, White and Hazard, who were at that time operating a wire works at the Falls of the Schuylkill River near Philadelphia, were faced with a major calamity when the Findlay-type chain bridge that their workers had to cross to reach their mill collapsed under a load of winter ice. They quickly replaced the fallen span with a foot bridge suspended from iron-wire catenaries. This was the first iron-wire suspension bridge in the world and it was the direct ancestor of such modern behemoths as the Golden Gate and Verrazano Narrows bridges.[10] Although their iron-wire bridge was only a temporary expedient, White and Hazard

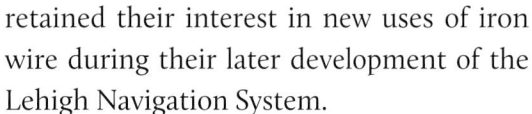

The iron bands used on the inclines worked well, but were limited in their capacity. They were attached to a "barney" car (on right), which pushed loaded cars up the slope. At the top of the incline the iron bands wound around large wooden wheels. Steam engines at the summit powered the system. (NCM/D&L)

retained their interest in new uses of iron wire during their later development of the Lehigh Navigation System.

During the 1820s and 1830s, European wire makers had developed iron-wire rope, a product that found many applications such as bridge cables and ship rigging. This new technology was brought to America in 1831 by Charles Ellet, a young canal engineer who had studied in France. A pamphlet Ellet wrote in 1839 on iron-wire suspension bridges stimulated the thinking of John A. Roebling, a young German immigrant who was working as a canal engineer in western Pennsylvania and Ohio. After many experiments, Roebling successfully produced iron-wire rope at an outdoor, hand-operated rope works in Saxonburg, Pennsylvania, in 1842. His product was first employed on inclined planes associated with the Pennsylvania Main Line Canal and later on that route's Allegheny Portage Railroad.

In 1844 one of Roebling's handmade iron-wire ropes was purchased by the Delaware and Hudson Canal Company, which wanted to try it out on an inclined plane of their gravity railroad between Carbondale and Honesdale, Pennsylvania. This caught the attention of White and Hazard, who, along with the Lehigh Coal and Navigation Company's chief engineer, Edwin A. Douglas, were at that time confronted with a major technical dilemma. They were about to begin construction of the company's first inclined-plane railroad and were concerned about the liabilities of using hemp ropes or iron chains on the planes.

Hemp ropes were expensive and quickly wore out, while iron chains such as were used on the inclined planes of New Jersey's Morris Canal had an unfortunate tendency to break while under strain, causing great damage. To eliminate the use of ropes and chains on their planes, White, Hazard, and Douglas had designed their inclines to use long flat iron straps that wound and unwound, like watch springs, on large drums powered by steam engines as they pulled loaded coal cars up and lowered empties. Although the iron bands were successful and functioned on the two planes of the back track of the Summit Hill to Mauch Chunk Gravity Railroad until 1932, on longer and steeper planes the bands would not work. After learning of Roebling's handmade iron-wire ropes, White, Hazard, and Douglas (who were aware of the virtues of iron-wire rope through their reading of European technical literature) ordered one in 1846. The wire rope was soon put in use on the chutes that loaded anthracite into the canal boats on the Lehigh Navigation at Mauch Chunk. The handmade rope wore well and had great strength, so the Lehigh Coal and Navigation Company ordered six additional wire ropes from Roebling.

Hazard and Douglas were determined to develop their own source of wire rope. Hazard sketched out a mechanically powered wire-rope machine that was based on the design of a "French bobbin machine" (for making lace) that he had seen on a trip to Europe many years earlier. As constructed by Douglas and his assistants, George A. and Robert H. Sayre, this machine consisted of a horizontally mounted wooden cone with spools of wire mounted on its wide end. As the cone was mechanically rotated, wire from the spools was unwound, twisted, and drawn through an opening in the cone's narrow end, producing wire rope. The machine was installed in a former water-powered grist mill at the foot of Race Street in Mauch Chunk. This pioneering wire-rope factory was the first of its kind to operate in the United States. During 1848–1849 John A. Roebling, who had come east to spin the suspension cables for the aqueducts of the Delaware and Hudson Canal Company, visited the Mauch Chunk wire-rope factory and sketched its machinery. Later he would build his own factory to make wire rope along the Delaware and Raritan Canal at Trenton, New Jersey.

Much of the wire used during the early days of the Mauch Chunk wire-rope factory was supplied by the Rodenbaugh and Stewart Company, whose mill was located along the Lehigh Navigation in what is now the Hugh Moore Park in Easton, Pennsylvania. In 1857 wire rope from the Mauch Chunk factory was taken to a site near Easton to form the basis of the suspension bridge that carried the towpath of the Lehigh Navigation from Island Park to the south shore.

Although the Mauch Chunk wire-rope factory was established to serve the needs of the Lehigh Coal and Navigation Company, it also became a profitable enterprise with many outside customers.

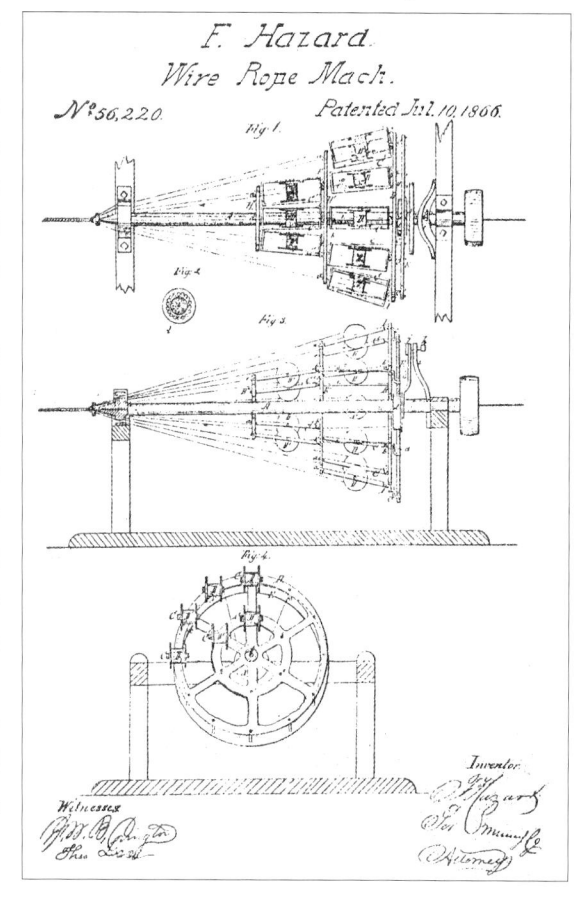

Fisher Hazard's machine to make wire rope, based on the French lace bobbin machine sketched by his father years earlier, used an intricate twisting pattern to create the rope. This drawing accompanied the 1866 patent. By that time, wire rope was widely used in mining and bridge building. (NCM/D&L)

W.W. Munsell & Co., History of Luzerne, Lackawanna, and Wyoming Counties, PA, 1786–1880.

Fisher Hazard, son of Erskine, in partnership with E. A. Douglas, leased the factory from the Lehigh Coal and Navigation Company in 1852 and managed it successfully for more than a decade. The Morris Canal became his biggest customer, purchasing wire ropes for use on its inclined planes. In 1866 Fisher Hazard obtained a patent for improvements to the wire-rope machine and established a wire mill to supply it with raw material. However, business reverses in his other enterprises caused him to incorporate the wire-rope factory as the Hazard Manufacturing Company in 1868 in order to sell it to Wilkes-Barre interests headed by Charles Parrish.

In 1870 the machinery of the wire-rope factory was transported to Wilkes-Barre and set up in a larger building. During the last decades of the nineteenth century the plant of the Hazard Manufacturing Company was greatly enlarged, and the company widened the range of its products to include electrical cable. In 1906 the plant had 700 employees. By the time of its incorporation into the Connecticut-based American Chain and Cable Company in 1927, it had grown to a complex of seventeen buildings containing over 400,00 square feet of manufacturing space. Between 1956 and 1957 this complex was abandoned and during 1991 it served as office space and housed light industry. The wire and wire-rope manufacturing machinery was transferred to a new structure at two nearby industrial parks in Luzerne County. Hazard brand wire rope is still manufactured in the Wyoming Valley by the Bridon American Corporation.[11]

Chapter Six

1850 TO 1860
Mineral Wealth

"If steam powered the industrial revolution, then it was coal that fueled it. Bringing those two elements together was the job of the railroads. Together, these two forces—coal and railroads—unshackled the world from waterways, which had been the source of both power and transportation for millennia."

Martin R. Karig, Coal Cars: The First Three Hundred Years

Coal replaced wood and waterpower as America's energy source with stunning speed. Anthracite allowed America's Industrial Revolution to build a literal full head of steam in only two decades. The result was a radical remaking of the landscape of life—economic, political, demographic, cultural, even the very use of the land itself—not only of eastern Pennsylvania, but of the entire country.

Anthracite + Iron = Manufacturing Revolution

In 1854, 45 percent of the iron produced in the United States was from anthracite-fired furnaces—303,000 tons. Bituminous-fired furnaces accounted for only 49,000 tons, while charcoal furnaces—located in remote areas of the country where coal was not available, and each producing only small amounts of iron—were still slightly ahead overall with annual output of 306,000 tons. "The coming of anthracite coal thus quickly assured American manufacturers for the first time of an abundant [and therefore cheap] domestic supply of iron," wrote A.D. Chandler in his seminal work, *The Visible Hand: The Management Revolution in American Business.* This led to a rapid, fundamental change in the processes of manufacturing in all types of products.

> Inexpensive iron and coal permitted the factory to spread quickly in the metalworking industries. Not only did the output of establishments making axes, scythes, hoes and plows increase, but for the first time the fabricating and assembling of interchangeable parts became widely used in making metal goods besides guns for the United States army. ... During the late 1840s manufacturers first began to use the technology of interchangeable parts in factories to produce newly invented machines, such as sewing machines and reapers. The need for specialized machinery in all these industries led to the creation almost overnight of the American machine tool industry.[1]

Hot-burning anthracite was the fuel of choice not only for metal-working industries, but also for virtually every type of manufacturing, because of the rapidly spreading use of stationary steam engines, which both heated and powered the new factories. Reliable delivery of the fuel came to mean that those factories could be located near population centers with abundant labor pools, all of which led to the rapid industrialization of New England, the Middle-Atlantic States and the Ohio Valley, and the upper midwest east of the Mississippi. By 1860, coal—at this point, almost entirely Pennsylvania anthracite coal—had completely eclipsed wood and waterpower as the nation's energy source. The ever-rising demand for anthracite led to major changes in eastern Pennsylvania's anthracite regions.

Coal Rides the Rails

During the 1850s and 1860s the size and nature of the anthracite-mining industry changed dramatically. Railroads began to penetrate from tidewater into the anthracite-coal regions, providing competition to the canals and navigation systems for transportation of anthracite. The Schuylkill region, already the most productive of the anthracite-mining areas, led the way when, in 1841, the Philadelphia and Reading Railroad was completed between Philadelphia and Pottsville. This helped to ensure the dominance of the Schuylkill region in Pennsylvania's coal trade for almost two decades. Transporting anthracite by rail was not only faster than by canal, it was also year-round. Before trains, coal shipments halted during the winter months, because the canals were closed and drained to prevent ice damage to the banks, locks, and canal structures. The railroads themselves became major consumers of anthracite, as the coal was the most efficient fuel for steam locomotives.

Asa Packer, circa 1860

The first railroad to compete directly with the Lehigh Coal and Navigation Company for the business of carrying coal from the coalfields down the Lehigh River to Allentown, Bethlehem, Easton, and ultimately well beyond, was the Lehigh Valley Railroad. Originally chartered in 1846 as the Delaware, Lehigh, Schuylkill and Susquehanna Railroad Company, it languished, construction barely started, until director Asa Packer determined to break LC&N's coal-transportation monopoly. Construction of the 46-mile-long railroad was begun in 1853 and completed in 1855. Numerous branch lines brought coal from the mines to the railroad. Junctions with New Jersey railroads connected the line to Philadelphia and the New York harbor area. Not only did this line enable year-round shipment of coal, it also introduced passenger and freight service, making possible major changes in the way of life of many along the route.

The completion of the Lehigh Valley Railroad between Easton and Wilkes-Barre in 1867, and the completion of the extension of the Lehigh Coal and Navigation Company's Lehigh and Susquehanna Railroad between Wilkes-Barre and Easton in 1869 created a new set of conditions allowing greatly increased production in both the Lehigh and Wyoming regions. Both of these railroad lines linked the northern, middle, and southern coalfields, an achievement that was unmatched by their primary rivals, the Delaware, Lackawanna and Western, and the Philadelphia and Reading railroads.

Asa Packer, 1805–1879

An effusive nineteenth-century biography of Asa Packer called him "one of the most conspicuously useful men in the great commonwealth of Pennsylvania." His seemingly boundless energy and business savvy also made him the richest man in the state. But Packer's success did not begin until he was nearly 30.

Packer was born in modest circumstances in Mystic, Connecticut, and as a youth received very little education. At 17, he walked from Mystic to Brooklyn Township, Susquehanna County, Pennsylvania, to apprentice himself to a cousin who was a carpenter. By 1833, married, a father, and a failed farmer, he contracted with the Lehigh Coal and Navigation Company to run a canal boat. As many boat captains did, Packer soon found that the small initial outlay to lease a canal boat could yield a significant cash income, so he contracted for a second boat and made his brother-in-law, James Blakslee, captain. With anthracite transport booming, Packer quit riding the canal himself (though he kept a financial interest in a string of boats), bought a general store (which he put Blakslee in charge of), and called up his earlier carpentry training to open a boatyard in Mauch Chunk.

These enterprises quickly made him a wealthy man by local standards, and his contracts for construction of the locks for the Upper Grand Section of the canal, completed in 1839, cemented his fortunes and his prominence in the community. In 1841 he began the first of two terms in the Pennsylvania General Assembly, 1841–42 and 1842–43. He later served two terms in Congress starting in 1853. Famously taciturn, Packer apparently never addressed either body, but his behind-the-scenes work while a state legislator is given the credit for the foundation of Carbon County in 1843. He accepted a five-year appointment as associate judge of Carbon County after serving in the legislature, and ever afterwards was known as Judge Packer.

Despite the fact that he had earned his wealth because of the Lehigh Navigation, there was no love lost between Packer and the LC&N, which he and many other competitors considered to have an unfair monopoly. He was one of the original investors in the Delaware, Lehigh, Schuylkill and Susquehanna Railroad. In 1851 he purchased most of its outstanding stock. Determined that his gamble would pay off, Packer hired Robert H. Sayre away from the LC&N, and entrusted his young chief engineer with the task of completing the now-renamed Lehigh Valley Railroad parallel to the Lehigh Navigation from Mauch Chunk to Easton. The year it was completed proved to be the high point of coal transport on the Lehigh Navigation; after 1855 the canal slowly but steadily lost anthracite tonnage to the LVRR.

Though he preferred to entrust the day-to-day management of his enterprises to Sayre and his other hand-picked managers, Packer remained firmly in control of the LVRR throughout his life. His later years were marked by philanthropy: he founded Lehigh University with a bequest of land and $500,000 in 1865, and later donated another $400,000 to build the library in memory of his daughter Lucy Packer Linderman. He ultimately left the University well over $1 million. In 1876, he purchased and donated the land for St. Luke's Hospital in Fountain Hill, and endowed $300,000 in his will. Trustees Robert Sayre and Elisha Packer Wilbur, Packer's nephew, private secretary, and ultimately his successor as LVRR president, used the endowment to greatly enlarge and staff the hospital.

The Packer Mansion, built in 1861, was willed to the borough of Mauch Chunk by his daughter, Mary Packer Cummings, and is a National Historic Landmark.

Going Deeper

Coalmining methods underwent a major evolution during the 1850s and 1860s. By 1860 almost all of the anthracite outcroppings that lay close to the surface had been exhausted. As the demand for anthracite increased, coal operators were forced to develop extensive and expensive underground mines. The adoption of more costly underground mining technology greatly benefitted the mining companies of the Wyoming region because their coal deposits were closer to the surface and were largely horizontal. These two geological factors greatly reduced their costs of underground mining in comparison to the expenses of underground anthracite mining in the Schuylkill and Lehigh regions, where the coal beds were deeper and pitched at extremely steep angles. Once direct railroad links between the Wyoming field and the eastern markets of Philadelphia and New York were completed, the lower costs of underground mining in the Wyoming field enabled it to assume supremacy in the production of anthracite lasting well into the twentieth century.[2]

Blasting Powder

When miners moved their operations underground to follow the anthracite deposits, it became necessary to use blasting powder. The Wapwallopen blasting powder mills in Luzerne County became an integral part of Pennsylvania's anthracite coalmining industry. As the demand for blasting powder increased, powder mills began to be established in the coal regions.

By 1840, Luzerne County led Pennsylvania in both the number of powder mills and in annual output.[3] The demand for blasting powder became so great that by the 1850s explosives for mining made up the largest proportion of powder manufacture nationwide.

The largest of the manufacturers, the DuPont Company, shipped its blasting powder into the anthracite coalmining regions via the canals, as mule-drawn boats were safer for moving explosive powder than spark-throwing steam locomotives. By 1859, the demand for blasting powder had become so great that DuPont purchased a powder mill complex along the Wapwallopen Creek in southwestern Luzerne County. Sited in Nescopeck Township, the nearest town of any importance was Berwick, on the opposite side of the Susquehanna River. Transportation for the manufactured powder was provided by the North Branch Canal and the Lackawanna and Bloomsburg Railroad, both of which ran along the north bank of the river.

DuPont's Wapwallopen mills produced prodigious amounts of blasting powder—almost 50,000 kegs of it in

Wapwallopen Creek cascading through the "Powder Hole" shows the waterpower that the DuPont company harnessed to make the explosive blasting powder that was essential to mining. (NCM/D&L)

General Paul A. Oliver purchased 600 acres of land on Laurel Run and opened the Oliver Powder Mills in 1873. Like Wapwallopen, it used waterpower as a safer source of energy than burning coal. In 1893, it turned out 1,000 kegs of blasting powder a day.
(NCM/D&L)

1864.[4] The mills remained the major supplier of blasting powder to the anthracite mining industry through the last half of the nineteenth century, shutting down in 1912 when DuPont consolidated its anthracite region operations at Moosic, Pennsylvania.[5]

Breaker Developments

A major innovation in the anthracite industry during the 1840 to 1860 period was the widespread adoption of mechanical devices to clean, size, and separate newly mined coal. With the increasing availability of anthracite, consumers were becoming more concerned with the quality of this fuel and less tolerant of dirt and slate in the coal they purchased. Customers also began to demand a variety of sizes of anthracite to meet their various needs. To solve this problem, anthracite-mining concerns began to build special facilities to clean and size anthracite before it was marketed. These facilities, which became known as breakers, started to be built in the 1840s after Joseph Battin, the superintendent of a coal-gas plant in Philadelphia, developed a mechanical crusher and screening device; he licensed his invention to Gideon Bast, who built the first breakers in the Schuylkill coalfields. By the 1860s use of coal breakers had become widespread in both the Wyoming and Lehigh coalfields.[6] A pernicious side effect of increasing numbers of breakers, however, was the great expansion of hazardous jobs for young boys.

A depiction of an early coal breaker from Eli Bowen's 1852 Pictorial Sketch-Book of Pennsylvania.

In this view of a breaker in the 1860s large piles of culm can be seen in the background. The breaker is small compared with the behemoths built later. (NCM/D&L)

Breakers also posed hazards for workers in the mines below. For much of the nineteenth century, breakers were constructed solely of wood, and were sited right over the top of the entrances of the mines. This practice led to the deadliest disaster in the anthracite regions, the 1869 fire at the Avondale mine in Plymouth Township, Luzerne County. Fire started in the wooden lining of the shaft and spread upward, igniting the wooden breaker and other buildings on the surface.*

With the burning breaker on top of the mine's only exit, there was no escape. One hundred and ten men and boys died in what was the worst mine disaster in the United States to that time.[7] The Pennsylvania State Legislature acted promptly and passed a mine safety law in 1870 that applied to all anthracite mines. Each mine had to have two entrances (owners of existing mines had to install a second entrance within four months), and all new breakers had to be located a "safe distance" from the shaft—but existing breakers over shafts were allowed to remain. The law included a number of other safety measures that mine owners contested. Death and injury rates declined substantially in the next decades.[8]

The "grandfathered" breakers over the mine shafts remained; only two years later, in 1871, a similar fire trapped and killed 24 miners in West Pittston. It was not until 1885 that Pennsylvania passed a law requiring breakers to be sited at least 200 feet from mine entrances.[9]

Rescuers are hoisting a man and a boy up the shaft in this image from Frank Leslie's Illustrated Newspaper, *September 25, 1869. These victims are not identified, but several fathers died with their sons in the fire.*

---

* Whether arson or accident was never convincingly proved, nor was the actual starting point of the conflagration. *Harper's Weekly, Frank Leslie's Illustrated Newspaper,* and *The New York Times* were among the national publications that sent reporters and artists to depict the rescue efforts and cover the Coroner's Jury hearings.

New Americans for a New America

Industrialization both spurred and was spurred on by a rapid increase in the immigrant population. This was a change in the human profile of the people of the Corridor, and the nation at large, that was no less dramatic than the alterations to the natural landscape.

Prior to 1830, virtually all the increase in population was internal, that is, Americans having American offspring. The census of 1830 revealed that 98.5 percent of the U.S. population was native-born. That picture began to change radically in the decades leading up to the Civil War.

Only about 8,400 people immigrated to the United States in 1820 (the total U.S. population was a little over 10 million that year), the vast majority of them almost certainly from the British Isles and Germany. Immigration totals gradually increased to 23,322 by 1830. But between 1831 and 1840, immigration more than quadrupled to 599,000. Of these, about 207,000 were Irish, who were fleeing the poverty, the religious, economic, and political oppression, and the laws against owning the land they farmed in their homeland. The balance were Germans (152,000), British (76,000), and French (46,000.) By 1840, the national population was just over 17 million, of whom 800,000 were foreign-born. Most economists agree that the stunning acceleration of the industry and economy of the United States would not have been possible without the rapid increase in population by immigration.

The 1840s were not only a decade of rapid industrial growth in the Corridor and the nation at large, they were also years of enormous economic and political upheaval in Europe. Between 1841 and 1850 immigration nearly tripled again, totaling 1,713,000 immigrants, including at least 781,000 Irish, 435,000 Germans, 267,000 British and 77,000 French immigrants. The devastating potato famine in Ireland, exacerbated by British refusal to apply any humanitarian relief, drove more than half the surviving population from the island, with many poor and desperate people essentially indenturing themselves to get passage to America. Meanwhile, the failed revolutions of 1848 on the Continent sent many intellectuals and political activists, as well as ordinary farmers and craftsmen, fleeing to the freedoms on the other side of the Atlantic. Bad times and poor conditions in Europe drove people out, while land, relatives, freedom, opportunity, and jobs in the U.S. lured them in.[10]

The first obvious sign of change was religious. The Irish immigrants were the first significant numbers of Catholics to arrive in America; they came in such numbers that the U.S. population fell from 95 percent Protestant in 1840 to 90 percent Protestant only ten years later.

Immigrants to the coal regions of Carbon and Luzerne counties were generally of two classes during the first twenty years of the anthracite-mining boom. The first were experienced Welsh and English miners, and the second were common laborers, also from the British Isles, along with some Germans. Early immigrants were almost always men, usually single, or younger married men who had left families behind. Thus, the "patches," as miners' communities were known, were often rough-and-tumble places in their early years. Eventually, however, wives and single women began to arrive, especially after 1850. With them, and their subsequent children, came the need for schools and churches and many of the patches slowly began to coalesce into the towns of the coal region. Even so, conditions in many patches were primitive, and the residents' lives were completely dependent on, and controlled by, the coal companies.[11]

Iron Mining Fever

The first anthracite-fueled iron furnaces in the Corridor used a mixture of local brown hematite and New Jersey magnetite. While the New Jersey ore was easily delivered via the Morris and Delaware canals and the Lehigh Navigation, getting the ore from the widely scattered local mines to the furnaces along the Lehigh River was a very difficult task prior to construction of railroads. Horse-drawn wagons were the only means. As iron companies expanded, building additional and larger furnaces, this became increasingly unsatisfactory. Massive wagon traffic jams clogged the roads to the furnaces, as described in the 1914 *History of Catasauqua*:

> Ore for the furnaces was hauled by heavy teams from various mines throughout the County. A line of teams nearly a mile in length was a customary sight. During rainy seasons, the "ore roads" became well nigh impassable.[12]

The situation was worst along the Lehigh River from Coplay downstream to Allentown, a five-mile stretch that contained a dozen blast furnaces by 1855. The only way to relieve the congestion was by building a railroad to the ore districts. The lead was taken by the Thomas Iron Company of Hokendauqua and the Crane Iron Company of Catasauqua, which together sought a charter for a railroad between Catasauqua and Fogelsville, where there were very productive

The Catasauqua and Fogelsville Railroad was a partnership of the Lehigh Crane and Thomas Iron companies to carry iron ore from mines in southwestern Lehigh County to their furnaces. The 1,100-foot-long trestle constructed in 1856 over the Jordan River valley near Guth's Station (present day Guthsville) was the longest iron bridge in the U.S. at the time. (M.S. Henry, History of the Lehigh Valley)

iron mines. After prolonged efforts, the railroad charter was secured, and the construction of the Catasauqua and Fogelsville Railroad began in the spring of 1856. It was extended to Lock Ridge in 1864, where the Thomas Iron Company was planning on building furnaces, and to the magnetite mines at Rittenhouse Gap along South Mountain in 1865. The East Pennsylvania Railroad, between Reading and Allentown, was completed in 1859, while the North Pennsylvania Railroad, whose route from Philadelphia to South Bethlehem was completed in 1857, traversed the ore-rich Saucon Valley.

Ore wharves were constructed at various locations along the railroads for loading ore onto railroad cars. The ore was hauled to the wharves on wagons pulled by teams of two, four, or six horses or mules, weighed, and then unloaded onto long heaps towering over the wharf. The ore wagons, which had removable slats in the bottom, were dragged up onto the heaps where the slats were pulled and the load thus deposited. Every day one or two cars were shifted onto the siding alongside the wharves, where loaders filled the cars by hand. The next day the cars were moved on out to the furnaces.

In Lower Macungie Township in west-central Lehigh County, limonite mining was intensive from 1855 through the end of the nineteenth century. More than 80 mines were in operation during this period to serve the nearby furnaces at Alburtis, East Macungie, and Emmaus. Ore was sent also to the furnaces of the large iron companies along the Lehigh River. Ore wharves were constructed along both the East Pennsylvania Railroad and the Catasauqua and Fogelsville Railroad rights-of-way.

The shipment of ore by railroad did not eliminate damage to local roadways; it simply changed the areas subject to destruction. The ore still had to be carted from the mines to the ore wharves. The situation in Lower Macungie was described by Oscar Penrose Knauss, publisher of *The Macungie Progress* and son of Aquila Knauss, manager of the ore wharf in East Macungie:

> The delivery of this ore, by strings of teams, every weekday throughout the year was an interesting sight. Roads were generally poor and this continual hard wear by heavy wagons made and kept them worse. In summer months especially the procession of ore trains raised clouds of dust that hung over the wagons. In winter the snow was ground up and creaked, making an eerie sound.[13]

In 1884 Macungie attorney E.R. Lichtenwallner, writing about Lower Macungie in the *History of the Counties of Lehigh and Carbon*, described the "golden age" of mining in the township:

> Within a few years previous to the "financial crash" of 1874 some very rich and valuable deposits of hematite iron-ore were uncovered in this township, and it seemed as if almost everybody who owned a tract of land, however small, had been seized with the mining fever. Leases were made, shafts sunk, and the "hidden treasure" sought for everywhere. Ore-washeries and smoke-stacks seemed to spring up throughout the township like mushrooms in a hot-bed, while the fires from the chimneys of two furnaces and a foundry ... lit up the night with their lurid flames. Although many beautiful farms were laid waste, the owners thereof reaped a rich harvest in the shape of royalties, and considered themselves amply compensated for the unsightly gaps made in their land ...[14]

When it rained the situation was even worse; the roads became seas of mud. The following account by Albert Ohl, taken from his *History of Upper Saucon Township*, describes conditions there:

> The hauling was paid by the ton, therefore overloading was often the rule. Nobody not having lived in this era can half imagine the commotion. Wagons often stuck in the mud up to the axle, drivers cracking the long snake whips, with a report of Revolver shots, drivers cursing to high heaven. This was certainly a hard life, but they helped each other out, but it cost a lot of horses and many broken down wagons, and broken harnesses, the saddlers trade was good in those days, each teamster liked to boast of the heavy loads he could haul, but it did not pay, bridges were broken down, roads got into terrible condition, etc.[15]

Thomas Iron Works at Hokendauqua in 1859. This furnace complex used rails not only to move materials about the plant but also to receive raw materials.
(M.S. Henry, History of the Lehigh Valley)

The iron mines at Ironton, in north-central Lehigh County, were so productive that a separate railroad was built to move their ore to market. These mines, opened in the early nineteenth century by Stephen Balliet to provide ore to the charcoal-fired Lehigh Furnace, were operated principally by the Thomas Iron Company from the 1850s until the end of World War I. To get ore from Ironton to their furnaces at Hokendauqua, the company contracted with Tinsley Jeter, a real-estate developer and entrepreneur who later founded Fountain Hill, to construct the Ironton Railroad. The line was chartered on June 3, 1859, and by May 24, 1860, the first loaded ore cars were transported from Ironton to the Thomas furnaces. In 1862 the Siegersville branch was added to reach the ore deposits in that vicinity. On February 1, 1882, the Thomas Iron Company acquired complete control of the Ironton Railroad and continued to realize profits from its operation, even after the ore deposits were depleted, by transporting portland cement from the mills at Coplay, Egypt, and Ormrod.[16]

Rapid Expansion of Iron Furnaces

By the end of the 1840s, businessmen and merchants in New York, Boston, and Philadelphia recognized the investment potential in anthracite ironmaking, that is, iron produced in anthracite-fueled furnaces. In many cases, they hired local experts in the iron and/or ore extraction industries to spearhead their enterprises. With the advent of the railroads, it became possible to locate these new furnaces away from the original power-and-transport lifelines, the canals.

Carbon Iron Company

The first furnace of the Carbon Iron Company, originally known as the Poco Anthracite Furnace, was built at Parryville by Bowman Brothers and Company in 1855. The site was on the Lehigh Canal at Lock No. 13, and just upstream from the mouth of Big, or Pohopoco Creek. The firm was incorporated as the Carbon Iron Company on August 15, 1857.

Carbon Furnace was located in Parryville along the canal at Lock 13. The locktender's house is just visible to the left of the covered bridge. (NCM/D&L)

The original water-powered blast machinery was replaced with steam equipment in 1857; production that year was 3,217 tons of iron, produced in just 41 weeks of operation. The expanding iron market of the 1860s led to construction of two more furnaces. Both of the new stacks had closed tops to capture waste gas to heat the stoves and generate steam for the blowing engines. By 1875, the three furnaces had a combined output of 30,000 tons a year.[17]

Lehigh Valley Iron Company

During the summer of 1853, veteran charcoal ironmaster Stephen Balliet organized the firm of Stephen Balliet & Company to build an anthracite furnace along the Lehigh River. This firm, consisting of the senior Balliet and his sons, Aaron and Stephen Balliet Jr., and his son-in-law,

Standing in front of the first furnace of the Lehigh Valley Iron Company is Valentine W. Weaver Jr. His father was superintendent after the company was reorganized as the Coplay Iron Company in 1878, and remained in the position until 1884.

The furnace complex had three furnaces in 1878 when it changed hands. Behind the buildings in this picturesque view was the Lehigh River and the Lehigh Valley Railroad; a junction with the Ironton Railroad was close by. The product was primarily foundry iron.

(Elizabeth Weaver Moatz)

Benjamin S. Levan, purchased a parcel of land from Daniel Schreiber on the west bank of the river just downstream from Schreiber's Ferry (the present site of Coplay), for the furnace site. The foundation for No. 1 stack was laid and frame workers' houses were built in the autumn of 1853.

Stephen Balliet Sr.'s death in January 1854 led to the firm's reorganization and incorporation as the Lehigh Valley Iron Company with Joseph Laubach and Lewis A. Buckley admitted as partners. Joseph Laubach, a prominent businessman and politician, was elected president and Benjamin S. Levan, who had managed the Lehigh Furnace for his father-in-law since 1832, was appointed superintendent; both served until 1878. This iron company lasted until the mid-1890s before being permanently shut down.[18]

Thomas Iron Company

The anthracite iron boom was well under way when the founders of the Thomas Iron Company first met in Mr. White's tavern in Center Square, Easton, on February 14, 1854. Of the eighteen men who attended this first organizational meeting, all but one subsequently subscribed the capital stock of the company. The founders were men of considerable talent who had already achieved distinction and financial success in other spheres of endeavor; most were interested in investing in the iron industry simply to increase their fortunes. Very little was accomplished at the first meeting except to name the company in honor of David Thomas, and to fix the capital at $200,000.

During the course of the next few months, additional meetings were held and an earlier decision to organize as "general partners" was abandoned when a special charter from the State of Pennsylvania was granted and approved on April 4, 1854. This charter remained in effect for the next 84 years before it was finally surrendered in June 1942, nearly 15 years after the last furnace was in blast. The first board of directors included E.A. Douglas, William H. Talcott, Ephraim Marsh, Peter S. Michler, John Drake, Samuel S. Chidsey and C.A. Luckenbach. David Thomas was made trustee of real estate and his son Samuel was appointed superintendent.

The Thomas Iron Company's first and main plant at Hokendauqua, as it appeared in 1904. It was along the Lehigh Valley Railroad. The dam was for Guard Lock No. 6 of the Lehigh Navigation, on the left bank of the river. (Thomas Iron Company)

For the site of their furnace, the organizers of the Thomas Iron Company authorized David Thomas to buy the Thomas Butz farm on the west bank of the Lehigh River about one mile upstream from Catasauqua. The company town subsequently laid out became the nucleus of

Samuel Thomas, 1827–1906

Though only 27 at the time of his appointment as superintendent of the Thomas Iron Company, Samuel Thomas was certainly well qualified for the position. From the age of 16 he had worked for his father at the Lehigh Crane Iron Works. Then, early in 1848 at the age of just 21, he was sent to Boonton, New Jersey, where he supervised the construction of and successfully blew in an anthracite furnace for the Boonton Iron Works. Later in the same year he returned to Catasauqua to assist his father in the construction of No. 4 and No. 5 furnaces. Samuel Thomas was destined to become one of the leading figures in the Thomas Iron Company, serving first as superintendent and then appointed a director and elected president on August 31, 1864. He continued as president for the next 23 years, finally resigning the office on September 22, 1887, so that he might fulfill his long-held ambition to build a modern iron works in Alabama. He remained a director of the Thomas Iron Company until his death at Catasauqua on February 21, 1906.

present-day Hokendauqua. The blowing engines, like those used in most Thomas-built furnaces, were among the largest available at the time. Furnace No. 1 was quickly constructed and blown in on June 3, 1855; No. 2 furnace was put into blast on October 27, 1855.

The following year, soon after completing construction of its first two furnaces, the Thomas Iron Company purchased the Richard Mine near Mount Hope, New Jersey. This highly productive mine contained large deposits of high-grade magnetic ores that were used in the Thomas furnaces. By 1891 the mine had become the largest producer of iron ore in New Jersey, supplying over 84,000 tons in that year alone. The company continued to operate the Richard Mine until 1923, when it sold the property to the Philadelphia and Reading Coal and Iron Company.

The first two furnaces of the Thomas works were considered models for other ironmakers to emulate and were chosen by the noted English metallurgist Dr. John Percy in 1861 as the example to illustrate the part of his pioneering metallurgical text that dealt with American anthracite ironmaking:

> But the principal manufacture must always cling to the Lehigh, Schuylkill and Lower Susquehanna valleys, in Pennsylvania, where ore is abundant, the coal near at hand, and the flux on the spot; where the whole land is a garden, and, therefore, food cheap and plentiful, and the great seaports not far off. For all these considerations, as well as for beauty, size, and convenience of build, and for its historic interest, the Thomas furnaces, which have been selected as illustrative of the American manufacture for this work, stand pre-eminent.[19]

Other contemporary commentators likewise considered the Thomas works among the finest in America. Professor J.P. Lesley attributed their success primarily to the capable management of David and Samuel Thomas:

> The consummate skill and long experience of the manager must no doubt avoid or redress the ordinary troubles of the process, but this immense production even from these first class stacks can be accounted for only by the enormous consumption of oxygen which they are allowed. It is a satisfactory evidence of the constancy and reliability of the chemical and mechanical laws at our command for making iron that the introducer and oldest producer of anthracite iron in America has not been superseded, but is still able to lead off the greatly enhanced production with these high figures. It is evidently no game of chance, but a trial of practical wisdom based on experience and insured by the improvement of all the means at the disposal of man. In a word here stands the demonstration that a large and well built crucible, properly stocked with good ores and properly blown with power to spare, must be a great and continued success.[20]

In 1857, No. 1 stack produced 9,731 tons of iron and No. 2 produced 8,366 tons, far more than the average furnace of that day. Later, four more furnaces were added to the Hokendauqua plant. Later acquisitions increased the number of furnaces operated by the Thomas Iron Company to eleven and, for a brief time, twelve. The Lock Ridge Iron Company was acquired on May 1, 1869; the Keystone Furnace, chartered as D. Runkle Company, was purchased by April 1, 1882, and the works of the Saucon Iron Company were purchased on December 13, 1884. The Lucy Furnace was operated under lease from March 22, 1886, until December 15, 1887.[21]

The Zinc Industry

Many innovations in the zinc industry were introduced in the Lehigh Valley, and for many years the area was a leading producer of zinc ore. The origin of the zinc industry in the Lehigh Valley can be traced to the 1845 discovery of zinc ore deposits in the Saucon Creek Valley. A Saucon Valley farmer, Jacob Ueberroth, noticed some strange mineral deposits on his land which he brought to the attention of a Bethlehem educator and chemist, William Theodore Roepper, who determined that the minerals Ueberroth had shown him were a zinc ore called calamine. Roepper interested Robert Earp, a Philadelphia businessman and a major stockholder in the Lehigh Coal and Navigation Company, in his discovery, and with Earp's backing a small-scale mining venture was begun on the Ueberroth farm. During 1846 nine tons of zinc ore were extracted. Since the United States had no zinc smelting industry, the ore was sent to England, where it found no buyers. Earp withdrew his financial backing, ending the venture.

While the Saucon Valley deposits remained fallow, other entrepreneurs were founding the modern American zinc industry in the adjacent state of New Jersey. In 1848 the Sussex Zinc and Copper Mining and Manufacturing Company was organized to exploit the zinc deposits found at Franklin in Sussex County. Zinc was soon used for many purposes, including the production of brass, paints, and roofing materials.

Samuel Wetherill, a chemist working for the New Jersey concern, developed a process that greatly facilitated the smelting and refining of zinc in 1852. Soon afterwards, he left his job and came to the Saucon Valley, hoping that he could successfully exploit its zinc ore deposits. By the end of 1852 he had obtained the lease of the Jacob Ueberroth farm and he had also begun developing a company to both mine and smelt zinc. Wetherill perfected his process for smelting and refining zinc after forming a partnership with a Philadelphia entrepreneur, Howard Gilbert.

Wetherill's process for creating zinc oxide, which was the basis for many other zinc products, began by placing a layer of anthracite coal on the grate bars of a furnace. It was ignited, and when it had reached a white heat a mixture of two parts crushed zinc ore to one part anthracite coal dust was placed on it. As the mixture was consumed, the heavy white smoke that it produced was blown upward by a fan into a tower, where it passed into muslin bags. The zinc oxide contained in the smoke condensed and adhered to the inside of the bags while the impurities passed through the mesh of the bags and were exhausted by a smoke stack.

In 1853 Wetherill and Gilbert organized the Pennsylvania and Lehigh Zinc Company and began construction of their zinc works on the south bank of the Lehigh River opposite the community of Bethlehem. The works cost over $85,000 to build. By the end of the year they were producing up to four tons per day and the company expanded its operations to Bethlehem, where it constructed water-powered paint and barrel mills along the Lehigh Navigation. A small community that became known as Wetherill, in honor of the firm's founder, soon grew up near the south-bank zinc works.[22] However, Wetherill and Gilbert had to sell control of their company in order to finance its expansion and by 1854 the Pennsylvania and Lehigh Zinc Company was controlled by a group of Quaker entrepreneurs from Philadelphia. Wetherill and Gilbert continued

Pennsylvania and Lehigh Zinc Company's works on the south bank of the Lehigh River in 1860. By then the founders, Samuel Wetherill and Howard Gilbert, had been forced out by their investors in Philadelphia, and Joseph Wharton was in charge. Wharton leased the works and ran them until 1863. Later this site became part of the Bethlehem Iron Company, and subsequently of Bethlehem Steel.

(M.S. Henry, History of the Lehigh Valley*)*

to operate the zinc works under contract to the new owners.

The Philadelphia entrepreneurs soon became concerned over the safety of their investment and dispatched one of their number, Joseph Wharton, to inspect the zinc works. Wharton found that Wetherill and Gilbert had increased production at the expense of product quality. Ordered to remain at Bethlehem to oversee the operation, Wharton soon clashed with Wetherill and Gilbert and by 1857 they were bought out and Joseph Wharton was placed in total charge of the Pennsylvania and Lehigh Zinc Company. In 1858 Wharton leased the zinc works from the other proprietors but within a year he was once again reduced to the position of manager. In 1859 he leased the works again and began construction of a new furnace capable of producing metallic zinc or spelter. He successfully operated the works, producing both zinc oxide and spelter until the end of his contract in 1863.[23] This was not the end of Wharton's role in Bethlehem industry; on the contrary, it freed him to be part of the early days of the most significant industrial and economic development in the city and the entire region—Bethlehem Iron Company.

Chapter Seven

1860 TO 1870
IRON HORSES AND IRON RAILS

"When I see the rapid strides that business has taken in this Valley for the past 10 years and think of the impetus that the improvements now in contemplation will do in the next 10, I predict a future for it that will surprise its most sanguine citizens."

Robert Sayre, in a letter to John Fritz, May 1860

Asa Packer had a problem. During the course of construction of Packer's Lehigh Valley Railroad, large quantities of rails were purchased from the Lackawanna Iron and Coal Company, located at what is now Scranton, Pennsylvania. It was one of the largest American manufacturers of rails, but its rails were of poor quality and, equally as important, Lackawanna Iron and Coal was controlled by Moses Taylor, a New York entrepreneur who was also the chief financial backer of the Delaware, Lackawanna and Western Railroad.

The DL&W Railroad was at that time being rapidly built across the Poconos to haul anthracite from the Lackawanna Valley to connecting railroads in New Jersey. Every purchase of Lackawanna rails by the Lehigh Valley Railroad was therefore, in effect, serving as a subsidy for a potential competitor. The managers of the Lehigh Valley Railroad had few options, since almost all domestic rails were, at best, of mediocre quality, and superior British rails were expensive due to high tariffs.[1]

Packer, the Lehigh Valley Railroad's owner and president, was an astute—not to say ruthless—businessman. When he saw a problem, he wanted a solution. The solution to this problem was provided by Robert H. Sayre.

Launching the Lehigh Valley Railroad

Packer had hired Sayre, then aged 28, away from the Lehigh Coal and Navigation Company in 1852, and charged him with constructing the Lehigh Valley Railroad from Mauch Chunk to Easton. Sayre completed the job, from the initial surveys of the terrain to the first train trip, in only three years. Sayre personally drove the locomotive on the leg from South Bethlehem (which did not actually exist at the time) to Easton. He not only built the railroad and entered into contracts to lease the necessary rolling stock, but he designed and supervised construction of the

The Lehigh Valley Railroad bridge over the Delaware, where it junctioned with the Central Railroad of New Jersey and the Belvidere Delaware Railroad. After the bi-level bridge was completed, anthracite coal could be transported by rail to Philadelphia and to New York harbor far more efficiently and in far greater quantities than by canal. The original two-level wooden bridge was replaced in 1876 with an iron bridge. Trains on the Lehigh Valley RR and Central of New Jersey crossed on the upper level, and Bel Del trains crossed on the lower level.

In this lithograph from M.S. Henry's 1860 History of the Lehigh Valley, *the entrance to the Morris Canal is under the bridge. The Bel Del tracks are close to the river, while the CNJ tracks curve around to the far right of the image.*

innovative bi-level bridge across the Delaware at Easton that linked the Lehigh Valley Railroad with the tracks of the Central Railroad of New Jersey and the Belvidere Delaware Railroad, thus giving his line access to Philadelphia and New York.

With the railroad temporarily completed, Packer appointed Robert Sayre its general superintendent as well as chief engineer, which effectively gave him operational control of the rail line. His authority and independence of action were also increased by Asa Packer's tendency to serve as an absentee owner who devoted the majority of his efforts to dealing with financiers in Philadelphia and New York.

Under Sayre's direction, the Lehigh Valley Railroad rapidly increased its traffic volume, and by 1858 Sayre had moved its general headquarters from Mauch Chunk to Bethlehem to be closer to the railroad's primary repair shops and its junction points with connecting railroads such as the North Pennsylvania Railroad and the Central Railroad of New Jersey, which provided outlets to Philadelphia and New York respectively.

Sayre's move to the new community of South Bethlehem, where he built himself a large brick house overlooking the Lehigh River and his train tracks, placed him near the center of the Lehigh Valley's rapidly developing iron industry.[2]

The Saucona Iron Works

In 1857 a Bethlehem merchant, Augustus Wolle, became interested in developing iron ore beds on the Gangewere farm, which was located at the present-day Saucon Valley Country Club. Wolle organized the Saucona Iron Company to exploit the ore in these deposits. Asa Packer was among the initial subscribers, and, similar to the arrangement with his railroad, he directed Sayre to take an active role in its affairs. Realizing that the nascent enterprise, if properly managed, could provide an answer to the Lehigh Valley Railroad's rail-source dilemma, Sayre used the railroad's financial resources to take effective management control of the Saucona Iron Company.[3]

The company was reorganized in 1858 as the Bethlehem Rolling Mill and Iron Company, a name which better reflected its intended purpose. Sayre influenced the decision to build the company's plant at the junction of the Lehigh Valley and North Pennsylvania railroads. The choice of this site meant that after the plant was built and placed in operation, its products could be readily shipped to market in New York, Philadelphia, and the anthracite regions of Pennsylvania. Sayre also selected the skilled ironmaster who was needed to design the plant and supervise the company's operations. At the inaugural board meeting of the Bethlehem Rolling Mill and Iron Company, the directors hired John Fritz as their general manager and superintendent.

John Fritz (1822–1913) was perhaps the most mechanically innovative of America's ironmasters. He had served since 1854 as the superintendent of the works of the Cambria Iron Company at Johnstown, Pennsylvania. In 1857 at Cambria, he developed an innovative "three-high" rail mill that produced, for the first time in the United States, wrought-iron railroad rails of uniformly high quality at an economical price.

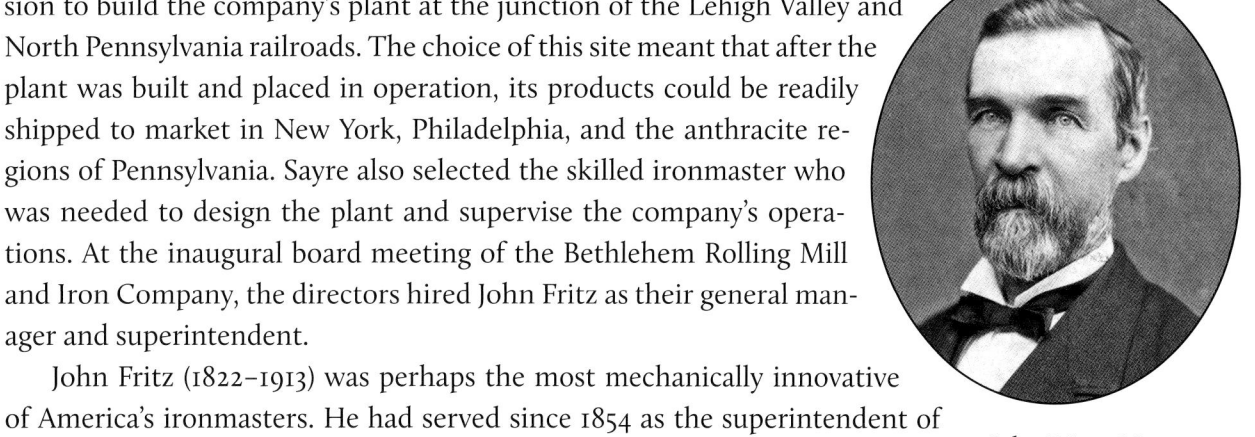

John Fritz, 1880

The 1858 patent model of the three-high rail mill developed by John Fritz and his brother George at Cambria is on display at the National Museum of Industrial History in Bethlehem.

On the side, three rolls can be seen. They rotated in opposite directions, enabling a pile to be placed between one set of rolls, reduced while still hot, and returned through the second set of rolls for further reduction before they had cooled. This procedure resulted in rails that were not likely to shatter, and was a very significant advantage over rails rolled in two-high mills, which could shatter in production and while in use.

(Glenn Koehler / Courtesy of the National Museum of Industrial History. All rights reserved.)

The small building in the center of the circa 1857 painting was where the original offices of Bethlehem Iron were located. Tracks of the Lehigh Valley Railroad ran through the plant then, and remained there until 1910, when the plant had increased exponentially, and they were moved closer to the Lehigh River.

(NCM/D&L)

Unlike the commonly used two-high mill, which was composed of only two sets of rolls, the three sets of rolls of the three-high mill enabled a red-hot wrought-iron pile to be completely rolled into a finished rail in one operation before it could cool and potentially shatter.*

The three-high rail mill was placed in successful operation on July 29, 1857, and John Fritz was granted a patent on his mechanical innovation on October 5, 1858. This patent became the basis for a pool that would eventually involve almost all of the major American rail mills.[4]

Under the terms of his contract with the Bethlehem Rolling Mill and Iron Company, Fritz was appointed general manager and superintendent of the company's works at a salary of $5,000 per annum, although the works were yet to be built. In addition, he received a total of 100 shares of the company's stock, to be paid in four annual installments, in return for his granting free use of the three-high rail mill patent.

Despite the ravages of the 1862 Lehigh River flood, work on the manufacturing facilities of the Bethlehem Rolling Mill and Iron Company proceeded rapidly. The entire plant was designed by John Fritz, who also supervised its construction. By the time the No. 1 Blast Furnace was placed in operation on January 4, 1863, the enterprise had been reorganized as the Bethlehem Iron Company. By July 27, the puddling furnace had begun to produce wrought-iron blooms for the rail-rolling mill, and by September 26, the Bethlehem Iron Company was making high-quality wrought-iron rails.

The financial support of the highly profitable Lehigh Valley Railroad enabled the Bethlehem Iron Company to expand its operations during the 1860s. By the end of 1863, the iron company's works had grown to include four stationary steam engines, a blast furnace, fourteen puddling furnaces, nine heating furnaces, a twenty-one inch (based on the diameter of the rolls) puddle train, and a twenty-one inch rail train. Bethlehem Iron Company's No. 2 Blast Furnace was constructed in 1867, and a year later a large foundry and machine shop complex was completed. To further increase its pig-iron production capacity, the company purchased an unused blast furnace from the Northampton Iron Company, which was located on an adjacent property.

---

* A "pile" is a stack of wrought-iron bars that were heated and hammered together and then fed into the rail mill.

Designated No. 3 Blast Furnace, its acquisition raised the company's ironmaking capacity to an annual total of 30,000 tons.[5]

The superior quality of its product soon gave the Bethlehem Iron Company a major share of the railroad-rail market in the east. However, a new product, Bessemer steel rails, began to appear in America during the 1860s. The superior durability of this British import attracted the attention of major American railroads. Although Bessemer steel rails were far costlier than wrought-iron rails, they lasted three times longer. As early as 1864, under Sayre's direction, the Lehigh Valley Railroad began to import Bessemer steel rails because its chief rival, the Lehigh Coal and Navigation Company, was extending its competing Lehigh and Susquehanna Railroad to parallel almost the entire route of the Lehigh Valley Railroad. The LC&N was using imported Bessemer steel rails, and Sayre feared that this innovation would greatly reduce the Lehigh and Susquehanna Railroad's maintenance costs and give it an economic advantage over the Lehigh Valley Railroad.

Robert Sayre, 1860

Sayre began to prod the Bethlehem Iron Company to investigate the production of Bessemer steel rails. Fritz, who had visited an experimental Bessemer steel works at Troy, New York, was opposed; the Troy plant had installed a small converter and the early results had been poor because of the presence of phosphorus, which was found in most American iron ores. Since a phosphorus level greater than 0.02 percent made steel produced in a Bessemer converter extremely brittle, Fritz felt that the Bessemer process was useless to most American ironmakers. During his tenure at Cambria, he had also witnessed William Kelly's singularly unsuccessful steelmaking experiments in western Pennsylvania. Kelly's experiments, similar in concept to the Bessemer process, had not resulted in a usable product, giving Fritz additional cause for his reluctance to commit Bethlehem to the construction of a steelmaking plant.

Fritz changed his mind after learning about a key discovery: that introducing spiegeleisen (ferro manganese) into the charge of a Bessemer would produce a steel suitable for rails. The Bethlehem Iron Company entered the Pneumatic Steel Association in 1867, propelling Fritz to the forefront of the efforts to create a viable Bessemer steel industry in the United States. He quickly absorbed the best available knowledge on the subject, to which he applied his mechanical-engineering genius.

With his brother George, general superintendent at Cambria, and Alexander Holley, who owned the rights to the Bessemer process, Fritz played a large role in designing the Pennsylvania Steel Company works at what is now Steelton, Pennsylvania. This plant, completed in 1867, was the first commercially successful Bessemer steel plant in America.

After returning from a trip in 1868 to examine steel works in England, France, Germany, and Austria, Fritz began work on the Bethlehem Iron Company's Bessemer steel plant, aided by Holley. Fritz's goal of making their plant the most efficient one in the United States slowed construction of the plant; it was not put into full operation until October 4, 1873.[6]

The Great Flood of 1862

On June 3, 1862, a heavy rain began to fall on eastern Pennsylvania. It rained all night, and all the next day. When the skies finally cleared on the morning of June 5, residents up and down the valley of the Lehigh saw the destruction wrought by the worst natural disaster ever to strike the counties of the Heritage Corridor. But the horror and devastation could not be blamed totally on Nature's uncontrollable forces. The damage that humans had wreaked on the landscape of the Lehigh River Valley in only thirty years by their breakneck quest for financial gain had increased the toll a thousand-fold.

The Lehigh Coal and Navigation Company, which bore the brunt of the financial and physical damage, particularly on its upper sections, had set the stage for the disaster by its staggering engineering feat of constructing the Upper Grand Section. The dams and locks had narrowed the Lehigh's natural river bed in many places, funneling water into tight channels that backed up the floodwaters and caused roaring currents.

Worse, the opening of the Lehigh Gorge to both coal mining and logging had resulted in steep hillsides that were denuded not only of trees but of almost all vegetation. This meant that rainwater simply ran down to the river, carrying soil and rocks with it. Worst of all, and the factor that multiplied the destruction incalculably, were the countless logs stored in the streams feeding the Lehigh itself above White Haven.

Author Joan Gilbert describes the onslaught of lumber in Gateway to the Coalfields:

> Harvested in winter, timber was floated down to sawmills on the Lehigh's tributaries after the ice melted. Lumber booms, strings of logs chained together and attached to the shore or stream bottom, were used to corral the cut trees ... With spring barely over when the storm began, the lumber booms were full. ... the rising water breached the small dams powering the stream-side sawmills and tore apart the lumber booms, hurtling hundreds of thousands of logs into the spate. Gathering momentum, they turned into battering rams, assaulting the Lehigh's long high dams. When one dam fell, the logs battered their way downstream to the next. By the time the last log escaped from its boom, the Upper Grand Section of the Lehigh Navigation was no longer recognizable. Nothing was immune to the onslaught.[7]

The devastation raced all the way down the river. People, mules, livestock, boats, buildings, bridges, roads, dams, walls, and trees were carried away. The eastward bend in the Lehigh at Allentown slowed the raging torrent somewhat, but there was flooding and destruction all the way to the Delaware. Every bridge that spanned the Lehigh was swept away; 200 boats were destroyed, with much loss of life, and eight lock houses lost. At least 150 people were killed.[8]

Joseph Levering, in his 1903 history of Bethlehem, called the Great Flood of 1862 "the most disastrous flood on record in the valley."[9] On the afternoon of June 4, 1862, the Lehigh River rose more rapidly than had ever been known before. Citing records maintained by the Bethlehem Bridge Company, Levering noted that the water gauge on the Lehigh River reached 20.5 feet, slightly above the 20-foot level of the Great Flood of January 1841. Upstream in Catasauqua, the high water was estimated to have reached 4.5 feet higher than the 1841 flood. The Lehigh River

Watercolor by Rufus Grider, an artist from Bethlehem, painted on June 5, 1862. Grider's note on the painting is: "The Flood from Nisky Hill at 8 o'clock A.M. a portion of the Beth^m Bridge just passing, the new Iron Works submerged — the Canal Bank covered." Logs and debris are surging downstream.

(Courtesy of the Moravian Archives, Bethlehem.)

eventually crested around noon on June 5, although the Delaware continued to rise, swollen by the very heavy runoff from tributaries entering the river basin.[10]

A little work published in 1863, called *Incidents of the Freshet on the Lehigh River, Sixth Month 4th and 5th, 1862*, describes what happened in low-lying Weissport:

> Weissport, owing to its low situation, suffered severely. It is thought that there was hardly a dwelling in the place escaped the effects of the water. Upon our first visit to it after the disaster, the scene of desolation it presented was appalling; lumber, wrecks of bridges, broken canal-boats, parts of carriages, etc., lay in endless confusion the length and breadth of the town. In its main streets lay canal-boats, parts of houses, and logs piled a story or more high for a long distance, effectually stopping all travel from it, and furnishing a sad memento of the overwhelming destruction. At the Fort Allen House the flood was on the bar-room floor several inches; the stabling and out houses attached to the hotel were all carried away. A resident of the place had taken much pains to furnish a correct account of the number of buildings destroyed. The whole number was eighty-nine, consisting of sixteen dwellings, thirteen

The 34-foot-high Tannery Dam in the aftermath of the 1862 flood. Perched atop the dam and lying in the foreground are great logs, some of the vast quantity of cut lumber that tore out all of the Upper Grand Section of the Navigation and wreaked havoc all the way down the Lehigh to Easton.

(NCM/D&L)

kitchens, thirty-seven stables, two barns, two blacksmith-shops, two slaughter-houses, two wagon-sheds, two built of brick, one school-house, one Methodist meeting-house, one saw-mill, one rolling mill, one foundry, one warehouse, and one carpenter-shop, coach-factory, cigar-shop, feed-store, shoe-shop, and tailor-shop. Four residents of the town were drowned.[11]

An empty grave?

A poignant monument to a Rittersville farmer and his two sons who died in the flood still stands in the cemetery of St. Peter's Lutheran Church on Hanover Avenue in Allentown. George Bickert had a farm on the banks of the Lehigh between Allentown and Bethlehem. He was 57 when he was lost and presumed dead on June 4, 1862. His sons, Joseph Henry and Milton, died as well. Their funeral was conducted by Rev. Daniel Brendle. His widow erected the monument, but it is not known if they are actually buried there, or were, like many of the flood's victims, never found.[12] The stone reads "In Memory of George Bickert and his Sons Henry & Milton Lost in the Flood of 1862"

(K.J.B. Ruch)

Making Tracks

The Lehigh Coal and Navigation Company quickly concluded that rebuilding the Upper Grand Section of the navigation was impractical, and concentrated its efforts on the original reach of the canal from Mauch Chunk to Easton.

> The devastation was so great between Mauch Chunk and Allentown that it involved a heavy outlay of money in lumber, iron, and other materials, and the labor of between two and three thousand men and five or six hundred horses and mules for nearly four months before navigation could be resumed. The first boat was loaded and started from Mauch Chunk the 29th day of the Ninth month, 1862. [13]

The rival Lehigh Valley Railroad recovered even faster, getting most of its trains rolling down the valley by the end of July. To replace the Upper Grand Section, LC&N got the state's permission to extend its Lehigh & Susquehanna Railroad through the Lehigh Gorge and then south and east to Easton.

As the line was being completed down the west side of the Lehigh through the gorge in 1868, the LC&N's head-to-head competition with the Lehigh Valley Railroad became even more intense. In 1863 Asa Packer directed Robert Sayre to extend the LVRR over the 1,700-foot-high Penobscot Mountain to reach Wilkes-Barre; though construction did not begin until 1866, the monumental task was completed in only three years. This gave Packer's railroad access to Wyoming Valley coal. While work was in progress, Packer also purchased the failing North Branch Canal, and Sayre used its towpath as a rail bed to carry the now-named Pennsylvania and New York Railroad to a junction with the Erie Railroad, and access to the Great Lakes port of Buffalo. Packer incorporated the line as a separate company so that if it failed, the loss would not damage the LVRR.

Meanwhile, the Lehigh and Susquehanna completed rail connections to the mines at Penn Haven, reached Easton, and crossed the Delaware to a junction with the Central Railroad of New Jersey (CNJ), allowing it access to the port of New York and the shipping lines that were increasingly carrying anthracite to port cities on the east coast, especially Boston. By 1870, because of the proliferation of railroads hauling anthracite across New Jersey, the amount of anthracite shipped through New York harbor exceeded that which passed through Philadelphia.

The Delaware, Lackawanna, and Western Railroad carried Wyoming Valley coal across New Jersey, also via a connection and an agreement with the CNJ. The Lehigh Valley Railroad leased the financially floundering Morris Canal solely to have access to its extensive waterfront properties in Jersey City. Though the railroad company left the canal operations more or less intact, the reality was that one railroad car could carry as much coal as five Morris Canal boats. The anthracite canals were able to survive, however, because their freight charges were lower and they were still able to deliver directly to customers along their routes, such as the iron furnaces at Catasauqua and Glendon. Only the Delaware Canal was free of the challenge of a railroad shadowing its waterway.

Robert Heysham Sayre, 1824–1907

Robert H. Sayre was born in 1824 on a farm in Columbia County, Pennsylvania. In 1829 his father, William H. Sayre Sr., was named weigh-lock master on the newly opened Lehigh Coal and Navigation Company canal in Mauch Chunk. Robert and his younger brother, William H. Jr., grew up in the midst of the great explosion of American civil engineering that was being carried out along the Lehigh by Canvass White, the engineer of the Erie, Lehigh, and Delaware & Raritan canals, and his protégé, Edwin A. Douglas, who engineered and directed the difficult building of the Upper Grand Section through the Lehigh Gorge.

At 17, Robert Sayre went to work with Douglas, first on the enlargement of the Morris Canal, then construction of the "back track" of the Gravity Railroad, the completion of the Ashley Planes, and the reconstruction of the Upper Grand after the flood of 1847. By 1850, Douglas's declining health put Sayre, aged 26, in charge of the LC&N's canal and rail operations. Two years later, Asa Packer hired Sayre to build the Lehigh Valley Railroad. Sayre convinced Packer to move the railroad's headquarters to the area that became South Bethlehem. Much of the development of Bethlehem South, as it was then known, into a flourishing multi-cultural industrial and educational community is due to this dynamic man and his immediate and extended family.

During the early 1860s, Robert Sayre was both running the Lehigh Valley Railroad and directing the development of Bethlehem Iron. In the former, he was assisted by his brother William, who had moved with his family and their father to South Bethlehem after losing their homes in Mauch Chunk in the 1862 flood. At the same time, the Sayres, devout Episcopalians, founded and built the Church of the Nativity, the second church in South Bethlehem, across the street from their homes.

In 1863, Packer directed Sayre to plan and construct a northbound branch of the LVRR to cross the 1,730-foot summit of Penobscot Mountain and link Mauch Chunk and Wilkes-Barre. From there Sayre built the newly formed Pennsylvania and New York Railroad on the former North Branch Canal towpath to a junction with the Erie Railroad in upstate New York. In 1871, Sayre leased the Morris Canal in order to secure control of its valuable waterfront properties on New York harbor, and built the Easton and Amboy Railroad to give the LVRR a rail route to ship anthracite directly to the port.

After Asa Packer's death in 1879, ownership of the LVRR fell to his sons Harry and Robert. Incompetent and given to drunkenness, both soon demonstrated their intense dislike of Sayre. By 1882, unable to tolerate their insults and constant challenges to his authority and expertise, Sayre resigned to become president and chief engineer of the South Pennsylvania Railroad. In 1885, after the deaths of the Packer brothers, Sayre returned to re-take charge of the Lehigh Valley Railroad; the same year, he became general superintendent of Bethlehem Iron.

Sayre had never completely relinquished his management role in the iron company, and had strongly supported every technical innovation proposed by John Fritz. In 1891, Sayre became vice president of the company, which he held until his retirement in 1899, while his son-in-law, Robert Linderman, ascended to the president's chair.

Robert Sayre outlived three of his four wives: Mary Evelyn, mother of his first nine children; Mary Bradford, the widowed daughter-in-law of Jefferson Davis; and Helen Augusta Packer, a cousin of Asa Packer. His fourth wife was Martha Finley Nevin, youngest sister of John Nevin, second rector of the Church of the Nativity. She was the mother of Francis Bowes (later Woodrow Wilson's son-in-law and a prominent U.S. diplomat), John Nevin, and Cecil Nevin Sayre (who died in infancy), all of whom were born when their father was in his 60s. Sayre died in his home across Wyandotte Street from Nativity Church in 1907, and is buried in Nisky Hill Cemetery.

Establishing direct rail links with both the Great Lakes and New York Harbor became a high priority during the 1870s and 1880s for the Lehigh Valley Railroad, while at the same time it expanded the extent of its coal lands and trackage in Pennsylvania's anthracite coal regions. Between 1868 and 1873 the management of the Lehigh Valley Railroad gained control of 32,000 acres of coal lands.[14]

As the nation approached its centennial, the anthracite-canal era entered its twilight years. Though several more generations of canallers and their mules and boats continued their slow, steady travels up and down the Lehigh, the Delaware, the Morris, the Delaware and Raritan, and the Schuykill canals, railroads became the way America moved freight and people.

Robert Sayre and family, in a photograph taken at his home on Wyandotte Street about 1900. Sayre is surrounded by his fourth wife, Martha Nevin Sayre to his right, and their sons Francis Bowes (who married Woodrow Wilson's daughter Jessie) and John Nevin; his eldest son, Robert H. Sayre Jr., and his daughter Elizabeth Sayre Cleaver. Son-in-law Robert Linderman (a grandson of Asa Packer) is second from left in the front row. To his left, his daughter Ruth May Linderman Frick was the last surviving person in the picture when she died in 1979. The little boy in the front center with the bow tie is Robert H. Sayre III, and the balding man in the back row with white mustache and beard is son-in-law Albert Cleaver, who was the chairman of the chemistry department at Lehigh University.

(NCM/D&L)

This view of Mauch Chunk and Mount Pisgah by popular illustrator Harry Fenn was published in Picturesque America *in the mid-1870s. The town has two very busy railroads, transporting primarily coal. The Central RR of New Jersey is on the left, and the Lehigh Valley RR is on the right, across the river. The Lehigh Navigation is below the LVRR right of way.*

Chapter Eight

1870 to 1880
Rise and Fall and Rise

"If it had not been for the interesting and exciting character of the business, but few men would have been willing to incur the trouble and anxiety, and to endure the physical labor and danger ... long enough to place the business on a commercial basis."

John Fritz, on making steel

The 1870s—in particular, the year 1873—brought several significant turning points in the industrial development of the Corridor. That year, the Lehigh Valley's anthracite iron furnaces reached the peak of their national importance, and the Bethlehem Iron Company began the production of steel. Two years earlier, David O. Saylor had taken out the first American patent for portland cement, and in 1873 began producing it in mass quantities in Coplay. Saylor cement won a medal for highest quality at the Centennial Exposition in Philadelphia in 1876. The Centennial year also saw the formation of the Bristol Improvement Company, formed to lure more manufacturing firms to the borough. Also, between 1870 and 1875, both the Lehigh Valley and Lehigh and Susquehanna railroads became part of larger systems that directly linked their trackage in the anthracite regions with New York harbor. The result of this was expansion of deep mining, especially in the Wyoming Valley.

Decline of the Anthracite Iron Industry

In 1873, the Bethlehem Iron Company began producing steel; that same year, the Lehigh Valley's anthracite iron furnaces reached the peak of their national importance. At that time, the Lehigh Valley possessed 47 operating furnaces that produced an aggregate total of 389,967 tons of pig iron, which made it the most important ironmaking region in America. However, within a decade the Lehigh Valley's anthracite iron industry had begun to show signs of gradual weakening that set it on a course of relative decline in comparison to other iron-producing areas. Many factors contributed to this, all of them related to geography and capital. After 1875, eastern Pennsylvania was no longer the best location to produce iron and most of the existing furnaces were too undercapitalized to modernize, a prerequisite for survival in an increasingly competitive environment.

When Harry Fenn was preparing "Canal-boats receiving Coal" for Picturesque America *in the early 1870s, the coal boats were still in heavy use, but no longer for supplying iron furnaces. The furnace companies had needs greater than could be delivered by boats, and were receiving coal by railroads. This is the foot of the inclined planes at Mauch Chunk, where coal came down chutes directly into boats.*

The chutes dominated the view from everywhere in the town, but this did not deter tourism—on the contrary, industrial tourism was a big draw. In this picture, the Lehigh and Susquehanna Railroad is seen under the planes. LC&N had extended this railroad from White Haven after the Upper Grand Section was destroyed by the 1862 flood.

So long as anthracite coal was the most viable mineral fuel for iron production, the Lehigh Valley was ideally situated. The canal builders who promoted the industry's development saw their dreams realized, and they prospered. The iron furnaces provided a huge new market for their coal, and canal traffic increased beyond their expectations. The ironmakers' use of New Jersey magnetic ores was a bonus. As the industry developed, canal boats that delivered coal to the New York market—boats that in earlier years had often returned empty—began making profitable westbound voyages carrying Morris County ores to the Lehigh Valley.

By the mid-1850s, however, the limitations of the canal system began to curtail the industry's growth. More and larger furnaces were built, and their demand for raw materials was significantly greater. It became increasingly difficult, and uneconomical, to lay in sufficient supplies of raw materials to keep the furnaces operating during the winter months when the canals were frozen over.

Railroads solved this problem. As early as 1854, the Lehigh Valley Iron Company at Coplay and the Thomas Iron Company's works at Hokendauqua were built along the proposed route of the Lehigh Valley Railroad, not along the Lehigh Canal. According to Samuel Thomas, there was considerable speculation as to which would be ready first, the Lehigh Valley Railroad or No. 1 furnace. As the furnace was completed on June 3, 1855, and the railroad was not officially opened until September 25 the same year, the first shipments of coal were delivered by the Lehigh Coal and Navigation Company to a dock across the slackwater pool behind the Hokendauqua dam.

For many years the railroads and the iron companies enjoyed a highly profitable symbiotic relationship—the railroads carrying the raw materials to the furnaces and consuming as much as 70 percent of the iron industry's product, pig iron, which was turned into rails, locomotives, and freight cars. By the 1870s, however, this relationship began to turn sour. The railroads, like the iron companies, sustained huge losses during the depression that followed the Panic of 1873. In their effort to regain profitability, the anthracite-coal carriers started to cooperate on setting their rates. Though this attempt at monopoly ultimately failed, the small merchant pig-iron producers were at the railroads' mercy and suffered accordingly.

The superiority of coke as a blast-furnace fuel led to many local ironmakers starting to mix coke with anthracite as early as the 1870s. However, the coke was produced in ovens hundreds of miles from the eastern furnaces and was transported by the same carriers that had driven up the delivered price of anthracite. Some companies, like the Bethlehem Iron Company, that had close ties to the railroads may have enjoyed more favorable rates than the small merchant pig-iron producers.

In addition to competition from the modern, efficient furnaces that were erected in the Pittsburgh District during the 1870s, the local industry suffered from overexpansion resulting from the iron boom during the Civil War. The huge profits made during the war and rapid expansion of the railroads in the post-war era led many "town fathers" to believe that a blast furnace was vital to the economic well-being of their communities. Often furnaces were built with locally subscribed funds and a lessee was sought to operate the plant. Such furnaces were often technologically deficient and many were operated for only a few years before being rebuilt or abandoned. The promoters of these plants generally expected a quick profit on their investment and were often reluctant to subscribe the additional funds required to make the venture a success.

The Panic of 1873 damaged but did not destroy the local industry, even though many of the smaller operators were forced to file for bankruptcy. Most were reorganized, with bankers taking control. The bankers in turn were unwilling to make the expenditures required to modernize. Their judgment appears to have been sound, because they realized the futility of trying to compete with the western furnaces.[1] One iron company, however, had the distinct advantage of having been founded by attorney, publisher, and banker William H. Ainey. His Lehigh Iron Company modernized as needed, and remained in business under his presidency until his death in 1907.

The Fates of Some Local Furnaces

Allentown Iron Company

Like Glendon Iron, the owners of the Allentown Iron Company were out-of-town businessmen seeking to invest in the early anthracite pig-iron boom. The Philadelphia shipping firm of Bevan and Humphries hired ironmaster Benjamin Perry to set up furnaces on the west bank of the Lehigh, on a site now occupied by the American Parkway Bridge. Perry put two small furnaces into blast by 1847. In 1851, the owners sold out to a group of Allentown and Philadelphia investors, who expanded the works to five furnaces. By 1880, the company was producing 60,000 tons of pig iron per year. The switch to a fuel mix of anthracite and coke later that decade boosted the works' output to 68,000 tons, out of only three furnaces.

The firm never recovered from the financial panic of 1893; the furnaces operated only sporadically for the rest of the decade. Around 1900, the Empire Steel and Iron Company of Catasauqua took over briefly, but soon concluded that modernizing the plant was too costly, and the furnaces were dismantled by 1904.

Furnaces of the Allentown Iron Company in 1889, photographed by William Mickley Weaver. Weaver was an outstanding amateur photographer and superintendent of the small furnace in Macungie. The Lehigh Valley Railroad, in the center of the photo, was completed in 1867, enabling anthracite deliveries by train.

Carbon Iron Company

Economic conditions during the 1870s led to the firm's reorganization in 1876 as the Carbon Iron and Pipe Company, Ltd. A foundry to produce cast-iron pipe was added to the works in 1883–84, and in the early 1890s a modernization program was undertaken. No. 1 stack was dismantled in 1893, and a year later No. 2 stack was scrapped. In 1893–94, No. 3 stack was completely rebuilt.

This comparatively modern rebuilt furnace had an annual capacity of 38,000 tons, more than the combined capacity of the earlier furnaces. Late in the 1890s the company switched from local and New Jersey ores to Great Lakes and foreign hematites; these richer ores, coupled with the introduction of coke instead of anthracite as the furnace fuel, increased the annual capacity to 40,000 tons.

The firm was reorganized in 1917 and renamed the Carbon Furnace Company. It was one of the last iron furnaces still in operation in the Corridor when its final cast was made in August of 1923.[2] A photograph of the furnace complex is on page 67.

Durham Furnace

Like the Crane and Thomas Iron companies, the iron works at Durham, in Bucks County, not only survived the 1870s, but thrived through a combination of prudent management and far-sighted, efficiency-boosting innovation. The firm of Edward Cooper and Abram S. Hewitt—the son and son-in-law of the famous inventor, manufacturer, and philanthropist Peter Cooper—purchased the furnace in 1864, but sold it the following year. They bought it back at a sheriff's sale in 1870, and proceeded to invest major time and money in upgrading the iron works. They demolished the stone anthracite-fueled furnaces, and built one sheet-iron stack with an annual capacity of over 20,000 tons. Edward Cooper, a brilliant engineer, developed a stove that raised the temperature of the blast to a new, higher degree; so-called Durham stoves were also cheaper to build and run, and dominated the industry until about 1910.

Cooper and Hewitt introduced innovative metallurgical practices by employing a full-time chemist in 1870. This soon paid extra dividends when analysis showed that Durham iron was low in phosphorus and thus suitable for sale to steel plants using Bessemer converters. By 1874, Durham pig iron was being shipped to such pioneering steel plants as the Cambria works in Johnstown, the Pennsylvania Steel Company near Harrisburg, and Winslow, Griswold and Co. of Troy, New York.

Durham's shipping facilities were greatly improved during the 1870s, when a connection was made across the Delaware River to the Belvidere and Delaware Division of the Pennsylvania Railroad. A steam-operated inclined plane was built from the company's dock on the Delaware Canal to a stock area near the vertical charging elevators. The plane and dock were linked to the railroad by means of a siding to a cable ferry across the river. Later this system was replaced by a connection to the Quakertown and Easton Railroad, which reached Durham from the west.

Durham Furnace, 1876. The new sheet-metal furnace, considered state of the art when it was installed, is visible behind the cast house with the tall, pointed windows. This was one of the most productive furnaces in the Lehigh Region, turning out over 20,000 tons annually of the low-phosphorus pig iron that was greatly in demand for use in steelmaking.

(NCM/D&L)

The Durham works remained in operation until shut down by the Panic of 1892. It was blown in again in 1899 by John Fackenthal. The following year the works were bought by a group of Philadelphia investors, but it operated only another eight years. By 1912, the site had been completely dismantled.[3]

Lehigh Valley Iron Company

The Lehigh Valley Iron Company survived the first few years of the depression of the 1870s, but by December of 1878 debts had mounted to the point that the furnaces were blown out and bankruptcy was declared. The firm was then taken over by bankers.

On June 18, 1879, it was reorganized as the Coplay Iron Company, Ltd., with Elisha P. Wilbur of Bethlehem as chairman and William H. Ainey of Allentown as secretary and treasurer. Valentine Weygandt Weaver, who earlier was employed by the Crane, the Thomas, and the Millerstown iron companies, and served as superintendent of the furnaces at Hokendauqua, Lock Ridge, and Macungie, became the first superintendent of the reorganized firm.

The Coplay Iron Company, formerly the Lehigh Valley Iron Company, circa 1889.

The Anna, Lehigh Valley Iron Company's engine No. 4, in about 1872. All the furnace complexes had narrow-gauge locomotives and cars to move slag to the slag dumps, and pig-iron bars from the cast house to storage areas.

Weaver remained superintendent until 1884, after which he was succeeded briefly by Harrison Bortz, who at the same time was also superintendent of the Lehigh Iron Company at Aineyville. Michael Fackenthal, previously secretary and superintendent of the Saucon Iron Company, then served as superintendent for a few years until succeeded in 1890 by Horace Boyd, who remained until the plant was liquidated.

The exact date of the company's demise is unknown but is most likely around 1894 or 1895. No. 1 stack was abandoned in 1892. Boyd, the last superintendent, was reemployed as superintendent of the Thomas Iron Company's Saucon Furnaces on July 1, 1895.[4] The site of the works, just downstream from the Coplay-Northampton bridge, was long occupied by the lumber yard of the now-defunct General Supply Company. No evidence of the furnaces remains.

The Saucon Iron Company, after it was acquired by Thomas Iron. The view is of the north side of the stacks in Hellertown. Furnaces were often constructed in close proximity to a railroad to facilitate delivery of raw materials and shipment of products, usually pig iron. (Thomas Iron Company)

Saucon Iron Company

Despite the comparatively modern design of the Saucon Iron Company's plant near Hellertown, the company had difficulty surviving the "dull times" following the Panic of 1873. It limped along, out of blast as often as it was in, until November 5, 1884, when the stockholders finally agreed to sell out to the Thomas Iron Company for $300,000. The sale price was established by Joseph Wharton, who offered to buy the plant himself for $300,000, "but would not stand in the way of the Thomas Iron Company if they made the same offer." The sale was consummated on December 13, 1884, and stacks No. 1 and No. 2 were redesignated No. 10 and No. 11 of the Thomas Iron Company.

After the takeover, Horace Boyd was appointed superintendent and the furnaces were put back into blast. When Boyd left in 1890 to serve as superintendent of the Coplay Iron Company, he was replaced by Lee S. Clymer, who supervised the modernization program undertaken in 1893–1894. The original pipe stoves were replaced with Durham-style stoves and No. 10 stack was raised to 75 feet. These improvements, along with the introduction of greater quantities of coke, raised the plant's capacity from 25,000 to 55,000 tons. The original blowing engines, housed in a massive stone-arch structure each with its own boiler plant on top, remained the only source of blast until 1916 when a more powerful, though secondhand, engine was installed to raise the pressure and volume of the blast. The Thomas Iron Company operated the works until 1921, when the plant was dismantled and scrapped.[5]

John Fritz, 1822–1913

Few men had a greater impact on the development of the American iron and steel industry than John Fritz. Born in Chester County in 1822, Fritz showed his aptitude for mechanical innovation at an early age. At the age of 23, his creative energy took him to the position of superintendent of the Moore and Hooven rolling mill in Norristown. His reputation grew rapidly in the burgeoning Pennsylvania anthracite-iron industry, and after a brief period at the Union Foundry in Catasauqua, he was hired to be general superintendent of the Cambria Iron Works in 1854. At Cambria, Fritz devised the revolutionary "three-high" process for rolling railroad rails, and patented it in 1858. This major breakthrough ended America's dependence on imported British rails, and sparked the rapid expansion of railroads in the United States.

Fritz was hired by Robert Sayre to be the chief superintendent of the Bethlehem Iron Company in 1860, and he made the company the premier producer of iron rails in the United States. He then led the company into the production of steel and armor plate, which resulted in Bethlehem Iron becoming the nation's leading defense contractor. He remained with the company until he reached the age of 70 in 1892; his birthday celebration that year, organized by the Engineers Club of New York, brought virtually every major industrialist, inventor, and investor in America to Bethlehem.

Known as a demanding, but fair and often kindly boss and an energetic mentor of Lehigh University engineering and metallurgy students, "Uncle John" Fritz was a trustee of Lehigh from its founding until his death, save for a gap of ten years from 1897 to 1907. In 1909, he donated the funds and personally supervised the design and construction of an up-to-date engineering laboratory for the university.

John Fritz in 1862, soon after he came to work for the Bethlehem Iron Company.

In 1891, he paid the entire cost and personally supervised the design, construction, and decoration of the first Methodist church in South Bethlehem, adjacent to the Lehigh campus, with the sole request that it be named the Fritz Memorial Methodist Episcopal Church in honor of his parents. His funeral in February 1913, attended by everyone from steel company laborers to politicians and titans of American steel, was possibly the largest such event in the history of Bethlehem. He is buried in a simple plot in Nisky Hill cemetery alongside his wife and their daughter who died in childhood.[6]

"If my friends (and I have a great many warm ones)—if my friends think they can come to Bethlehem and have a dinner on my seventieth birthday, and can have a good time in so doing, I ought not and will not stand in the way," the famously modest John Fritz is said to have told the organizers of the celebration for his 70th birthday. Fritz and his "great many warm" friends posed for this photograph on September 29, 1892, the day after the lavish dinner. (NCM/D&L)

Steelmaking Begins in Bethlehem

In many ways, the Bessemer steel plant that John Fritz designed for the Bethlehem Iron Company can be considered the first serious attempt to integrate production of both steel and rails. Robert W. Hunt, a pioneering metallurgist, chemist, and mechanical engineer who was involved in some of the earliest attempts to create a Bessemer steel plant in America, described Fritz's plant:

> He arranged his melting-house, engine room, converting-room, blooming and rail mills, all in one grand building, under one roof, and without any partition walls. He placed his cupolas on the ground and hoisted the melted iron on a hydraulic lift and then poured it into the converters. The spiegel is also hoisted and poured into the vessels ... Instead of depending upon friction to drive the rollers of the tables, Mr. Fritz put in a pair of small reversing engines.[7]

The "one grand building" Hunt describes was Fritz's brilliant cruciform building, the partial ruins of which still stand. The 1873 *Guide Book of the Lehigh Valley Railroad* informed passengers that

> The new building now erecting for the manufacture of iron and steel will be, it is said, the largest in this country and one of the largest in existence anywhere. It will be 105 feet wide

Bethlehem artist Rufus Grider painted this watercolor of the works of the Bethlehem Iron Company in 1874. The previous year John Fritz's ground-breaking design for a facility that put several processes into one large double cruciform building was completed. The furnaces were close by, and molten iron could easily be transported to the new complex to be converted into steel in one of four Bessemer converters, then rolled into rails or merchant bar stock—steel bars that were sold to outside customers.

Grider always included a date and some information on his sketches and paintings. His notation at the bottom of this one tells us that it is the "same scene from the same place as the former, in May 1874—or 22 [sic] years after the 1st was taken." By "the former," Grider was referring to the painting (p. 79) he had made 12 years earlier of the great flood of 1862.

(Courtesy of Moravian Archives, Bethlehem)

A Bessemer converter being blown at Bethlehem Iron in 1886. These converters were at the heart of the new direction Bethlehem Iron was taking under John Fritz's management.

Bethlehem's first Bessemer unit was not put into use until late 1873, because Fritz was not satisfied with the quality of steel being produced by earlier converters elsewhere.

The photograph is from the collections held by the National Canal Museum/D&L Corridor. It is by Louis Comfort Tiffany, and was taken during a two-week cruise on the Delaware and Lehigh Canals in June 1886, starting at Bristol and ending at Mauch Chunk, aboard a scow converted to a houseboat. Tiffany and a group of friends traveled up the canals to celebrate his engagement to Louise Knox, daughter of the president of Lafayette College. The cruise on the "Molly-Polly-Chunker" was documented in photographs and a daily log.

spanned by an iron and slate roof without supporting columns. It is 30 feet high to the eaves and is in the shape of a double cross of which the long arm [or main building] is 941 feet and the short arms 140 feet each, making the area covered 1493 by 105 feet. This is only surpassed by the mill at Creuzot in France, which consists of three buildings 60 by 1400 feet each.

The steel works will start with a capacity of about 600 tons of rails per week, planned and arranged for a threefold increase of the same. There will be three trains of rolls, 24, 26 and 30 inch diameters, driven by two condensing-engines of 48 and 56 inches diameter cylinders, of 46 and 48 inches stroke.

The mill will be remarkable not only for its enormous size and capacity, but for the many new labor saving conveniences introduced.

The iron work for the building as well as the machinery was all made at the Company's shops and foundry.

Bethlehem Iron at that time employed about 700 men and produced 30,000 tons of iron a year, two-thirds of which was rolled into rails on the site. The steel plant of the Bethlehem Iron Company was the tenth American Bessemer works to begin production.* By 1878–1879 it produced over 78,697 tons of steel, a figure exceeded only by the 84,356 tons produced by the Cambria Iron Company and the 95,475 tons produced by the Carnegie group's new Edgar Thomson Steel Company of Braddock, near Pittsburgh.[8] Bethlehem was thus one of the leading steelmakers in a competitive market with no single plant dominating the field. However, the production leadership that the Edgar Thomson works had achieved in 1875–79 was a harbinger of its later dominance.[9]

The Lehigh Tube and Coil Works of Albright, Son & Co. in Allentown were an example of a secondary industry making iron products for other industrial users. Albright and Son advertised that they were "dealers in wrought-iron pipe and fittings, brass and iron works of all kinds on hand and made to order. Leather belting, vulcanized rubber belting, hose etc." and many other items.

Many other entrepreneurs opened shops to use the now-abundant and affordable iron from local furnaces. Machine shops, rolling mills, foundries, boiler makers, and makers of industrial tools, machinery, parts of all kinds for railroads, and agricultural equipment became established along the railroads and near furnaces.

The pace of invention and innovation increased rapidly, as iron was no longer scarce and expensive.

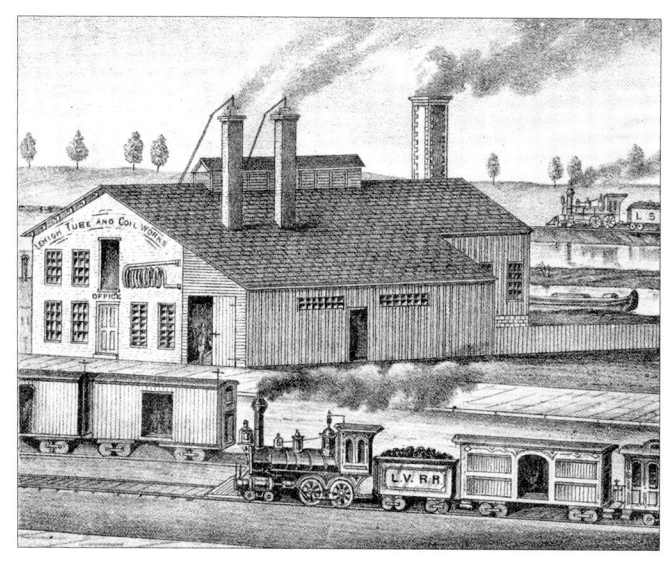

(Lehigh County 1876 Atlas)

---

* Steel is iron that has been chemically altered by high temperature and an injection of air (in modern steelmaking, oxygen alone) to reduce the carbon content. Existing pig iron furnaces made the basic material. The molten iron was then transferred to Bessemer converters and, later, to open-hearth furnaces, electric-arc furnaces, or basic-oxygen furnaces, where its chemical makeup was changed to steel.

Changes in the Anthracite Mining Industry

The Lehigh Valley and the Lehigh and Susquehanna railroads both became part of systems that developed through-routes connecting the anthracite coalfields and New York Harbor between 1871 and 1875. The Lehigh and Susquehanna reached the seaport when it was leased by the cash-starved Lehigh Coal and Navigation Company to the Central Railroad of New Jersey in 1871, while the Lehigh Valley Railroad completed its own line across New Jersey in 1875. This opened a direct route between the anthracite regions and the major market in the nation.

The Central Railroad of New Jersey acquired LC&N's 5,000 acres of Wyoming Valley coal lands and capitalized the Lehigh and Wilkes-Barre Canal Company to exploit them. The Lehigh Valley Railroad, which had earlier acquired substantial coal reserves in the Lehigh region through its mergers with the Beaver Meadow and Hazleton companies, purchased and developed extensive anthracite coal properties in the Wyoming region. Between 1870 and 1880 the Lehigh Valley Railroad purchased 3,166 acres of prime coal lands in this region, which were operated by its subsidiary, Lehigh Valley Coal Company, after 1875.

The added expenses of deep mining were a principal factor behind the consolidation by railroads of almost all of the major anthracite coal-mining properties in both the Lehigh and Wyoming regions; by 1880 this process was almost completed. Mining in the northern coal fields was relatively easier and cheaper than in the southern and western fields because Wyoming Valley coal lay in horizontal beds, rather than the vertical and diagonal, folded and faulted seams found in Carbon and Schuylkill mines. Once direct railroad links between the Wyoming Valley and the eastern markets of Philadelphia and New York were completed, the lower costs of underground mining there established the northern coal fields as the most productive anthracite mines, a position that lasted well into the twentieth century.[10]

The Cement Industry

Oddly enough, but like much of the rest of the industrial development of the Corridor, the roots of the cement industry lie in the construction of the canals. Hydraulic cement—which hardens under water—was necessary for building the locks of both the Lehigh Navigation and the rebuilt Delaware Division Canal. This was among the most critical technological innovations to be developed during America's canal era.

Hydraulic cement made possible long, large locks that otherwise would have had to be built of wood, which would not have had comparable structural strength, and would have rotted within a few years—exactly what happened to the original Delaware Division locks. Although it had been used in Europe since the Roman Empire, hydraulic cement was not manufactured in America until the building of the Erie Canal (1817–1825) when young American engineer Canvass White returned from England in 1817 with the secret of its manufacture.

Under White's direction, in 1826 an existing grist mill near Lehigh Gap was set to work grinding local argillaceous limestone. The ground rock was then burned in a lime kiln to produce hydraulic cement. The cement mill near Lehigh Gap operated between 1826 and 1830. A second cement mill was built in 1830 at Siegfried's Bridge, now Northampton borough, to supply the Lehigh Coal and Navigation Company.

Cement has been continuously manufactured in the Lehigh Valley since the founding of these two early mills. Until the end of the Civil War in 1865, almost all of the cement that was manufactured in the Lehigh Valley was used to build and repair the Lehigh Navigation and the Delaware Canal.

Hydraulic cement was also manufactured on a smaller scale in Bucks County. Clinker from Holland Township, three miles upstream across the Delaware River in New Jersey, was shipped to Narrowsville to be ground at Samuel Rufe's grist mill, supplying much of the hydraulic cement used in the Delaware Division Canal. Limestone was quarried and burned at Asher Ely's quarries in Solebury Township, two miles north of New Hope on the Primrose Creek. The clinker was ground at Phillips Mill. In 1829 this operation was producing 800 bushels of hydraulic cement a month but it ceased after 1833.[11]

Perfect for Portland Cement

Geological surveys made during the 1860s disclosed that the rock in many areas of the Lehigh Valley had unique properties that made it extremely suitable for making cement. This type of rock, called the Jacksonburg formation, combined chalk, clay, shale, and limestone. With the exception of an area around Boulogne, France, this type of rock is unique to the Lehigh Valley and adjacent areas of Berks County, Pennsylvania, and Warren County, New Jersey. It is found in a long belt between sedimentary deposits of limestone and shale, and is known locally as "cement rock" or "bastard" limestone. Large quarries, some abandoned and holding water, have been excavated the length of the Jacksonburg formation.

The cement industry, not just in the Lehigh Valley but in North America, was transformed during the 1860s and 1870s by the activities of David O. Saylor. In 1866 at Coplay, he started an enterprise that would later become the Coplay Cement Company. At first, Saylor's concern manufactured hydraulic cement; then, after he learned of the superior nature of British portland cement he began to experiment with making it. Unlike hydraulic cement, portland cement has a definite composition that is produced by mixing argillaceous limestone with other materials such as chalk, clay, and shale. This mixture is burned at a high temperature to produce portland cement. Portland cement is uniform in quality and when hardened it possesses a greater tensile strength than hydraulic cement.

David O. Saylor (1827-1884)

The Jacksonburg formation happened to possess all of the necessary ingredients for making portland cement. After several failures, by 1871 Saylor had produced the first true portland cement to be manufactured in North America, for which he was granted a patent. Saylor was assisted in his work by a young geologist and chemist, John W. Eckert, who

Coplay Cement about 1880. During the 1850s, construction of the Lehigh Valley Railroad tracks (in the foreground) uncovered the limestone deposits that allowed David Saylor to develop portland cement.

had recently graduated from Lehigh University. Eckert's employment by Saylor was considered by later historians to be the first step in the scientific progress of the portland cement industry in the United States. The results of Saylor and Eckert's work won first prize at the Philadelphia Centennial exposition of 1876, and the American portland cement industry was established.

The process developed by Saylor and refined by Eckert began by grinding rock from the Jacksonburg formation into a powder that was compressed into bricks. These were then burned in vertical dome kilns, using a design that had been imported from England. The result was clinker, which was reground into portland cement.

Saylor's product soon won wide recognition. In 1878 it was used in the construction of the Eads Jetties at the mouth of the Mississippi River. The jetties lined a two-and-a-half-mile channel through the mouth of the Mississippi River that connected the port of New Orleans with the Gulf of Mexico. Silt from the river had made the channel too shallow for the larger ships that became common in the second half of the nineteenth century. The silting-up had a severe impact on shipping revenues and other commerce in New Orleans. The jetties were built of large lumber poles with "mattresses" of woven willow branches and rocks fastened between them and topped with huge concrete blocks, mixed and formed on the site. By narrowing the channel, the jetties forced the river to flow through more quickly, which scoured out the silt and deepened the channel to 30 feet.

The Eads jetties were an important innovation, because it was the first time the principles of water movement had been incorporated into the design of a structure intended to use that movement to achieve a specific goal—in this case to maintain a channel 30 feet deep.

The economic impact of the jetties was enormous—exports from the port of New Orleans increased by 2,600 percent, and it became the second busiest harbor in the United States after New York.

The manufacture of portland cement in the Lehigh Valley was furthered when entrepreneur José de Navarro purchased patent rights to the newly developed rotary kiln in 1885. He and sons Alfonso and Antonio incorporated their Keystone Cement Company in Coplay in 1889, and began production late the same year. The name of the highly successful company changed to Atlas Cement in 1891, and expanded across the Lehigh River in 1896 to a site that became part of Northampton Borough. By 1899 there were 27 Atlas rotary kilns in operation.[12]

Another innovation that greatly aided the development of the cement industry was the invention of the Gates crusher, which speeded the preparation of rock for the kiln. By the end of the nineteenth century and during the early twentieth century the Lehigh Valley was the nation's primary cement producer, providing cement for concrete used in construction projects across the country and for the locks of the Panama Canal.[13]

Zinc in the Saucon Valley

The Civil War brought great prosperity to the Lehigh Zinc Company, since zinc was a necessary component of metallic shell cases. In 1865 Lehigh Zinc began rolling zinc sheets, which became widely adopted as a roofing material. The community known as Wetherill was organized as the borough of South Bethlehem in that year. James MacMahon, the superintendent of the zinc works, became the first chief burgess of the new town.

By 1870 the zinc ore deposits of the Saucon Valley were being extracted at five mines, all centered on the village of Friedensville. Four of the mines were open-pit strippings; one, the New Hartman Mine, was a deep-shaft operation. The largest and for many years the most important was the original Ueberroth mine, but the Hartman Mine, New Hartman Mine, Correll Mine, and the Triangle or Three-Cornered Mine were also major producers of zinc ore. All of them were operated by the Lehigh Zinc Company until 1876 when the lease of the Correll Mine was transferred to the Bergen Point Mining Company of New Jersey. In 1881 Franklin Osgood, part owner of the Correll or Saucon Mine approximately 1,800 feet south of the Ueberroth Mine, purchased the properties of the Lehigh Zinc Company and unified them under the corporate designation "Friedensville Zinc Company."[14]

The zinc ore bodies were found in a formation of fractured limestone that allowed water to pass easily into the mines. Until the miners reached a level forty feet below the surface this seepage was not a problem, but once they had passed that level the water flow became a torrent, halting mining operations. Company engineer John West designed a mammoth steam-powered pump designed to be installed at the Ueberroth Mine in 1869. It was expected to de-water all the interconnected mines. The Cornish beam pumping engine was manufactured in Philadelphia and shipped in sections to Center Valley via the North Pennsylvania Railroad, where the parts were loaded onto large wagons and hauled to the mine by teams of mules straining against each other to counteract the weight and act as a brake. Each flywheel had a diameter of 35 feet and weighed 92

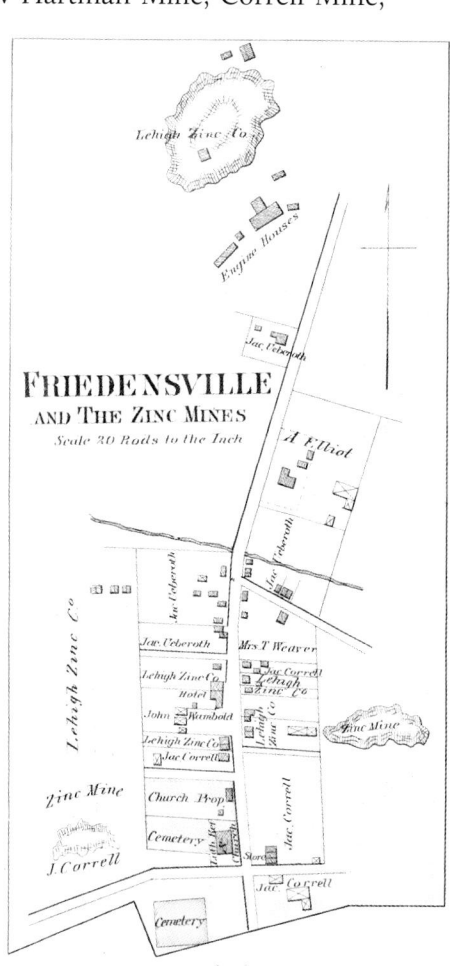

(*Lehigh County 1876 Atlas*)

Four stereoscopic views from the Library of Congress published by M.A. Kleckner, Allentown, dating from 1872–75. Above left, the pump house for the Cornish beam engine known as The President stands in the background. Above right is the engine house for the pumping engine. The photo was taken from the mine pit and also shows the entrance to the Trotter vein. Below are views of the dangerous outdoor working conditions in the mines, which went deep below the surface in fractured limestone bedrock. The underground flow of water became very strong at 40 feet, requiring pumping when the shafts reached that depth.

tons; the walking beams each weighed 44 tons; the connecting rods, 44 feet long, weighed 8 tons each; and the main flywheel shaft weighed 18 tons.

The pumping engine was assembled at the Ueberroth Mine; when it was completed in 1872 it was the largest of its type to have been built in North America, and believed to be the largest in the world, capable of pumping 12,000 gallons per minute from a depth of 300 feet. It was placed in operation on January 20, 1872. It was called "The President," a name that may be associated with its formal dedication at which President Ulysses S. Grant was to have officiated. Grant, however, never reached Friedensville. While Washington D.C. newspapers reported he was in "the wilds of Pennsylvania," stories circulating in Friedensville (most likely apocryphal) suggested that he had stopped to see a friend on the way, had enjoyed a few too many drinks, and forgotten the dedication of the pump.

Although technically brilliant, "The President" was an economic failure. The high cost of operating it raised the price of mining Saucon Valley zinc ore to four dollars per ton, while competing ores from Franklin, New Jersey, could be mined for only seventy-five cents per ton. Within a decade the pump was shut down, a decision that was popular among the farmers of the area, whose wells and springs had gone dry.

The Ueberroth Mine produced a small amount of ore in 1883 and again in 1886 when its pumps were operated to lower water in the other mines. The New Hartman and Correll mines were drowned in 1892 and attempts to lower the water level with pumps on rafts in the Ueberroth Mine were unsuccessful.

By 1893, the last mine had shut down, and zinc-mining operations ceased in the Saucon Valley until the mid-twentieth century. Total production by 1894 equaled 800,000 tons of ore averaging 30 percent zinc. By mine, this most likely would have been: Correll 100,000 tons; Old Hartman 200,000; Ueberroth 450,000; and Triangle (Three-Cornered) 50,000.[15]

Development of Bristol as an Industrial Center

While the counties to the north were undergoing massive changes, the area of Bucks County adjacent to the Delaware Canal for the most part remained decidedly rural, with market gardening a significant commercial activity along the upper section of the canal. Most of the businesses that grew up in that area were related to the commerce of the canal. There were several boat yards, most notably at Uhlerstown and Yardley, and numerous coal yards, though most of them north of New Hope were small operations. Stores, hotels, mule stables, blacksmiths, and grist mills were the most common commercial establishments. Stone quarries, most notably at Lumberville, were the heaviest industry along the canal.

South of New Hope the picture was different. The largest community was Bristol, which had grown and prospered during much of the canal era as a transshipment point for anthracite and other goods coming down the canal. The opening of the Delaware and Raritan Canal in 1840, and the 1848 outlet lock at New Hope that made it possible for boats to leave the Delaware Canal, cross the river, and head for New York Harbor via the new waterway, siphoned away a good

Above left, canal boats under construction at the boatyard at Upper Black Eddy, photographed by Louis Comfort Tiffany in 1886. Above right, a stone boat is being loaded with hard sandstone blocks at a quarry south of Lumberville.

Above, a coal boat is being unloaded at Leedom's Yardley coal yard. The two sections of the boat have been unhinged, and the front is being emptied first.

Left, a loaded boat, "Star of Bethlehem," is locking down at Lodi. This lock was never enlarged. It retains its original dimensions of 11 ft wide and 95 ft long.

deal of trade. Nevertheless, Bristol continued to prosper until about 1855, when it was seriously affected by the national economic downturn.

The transportation demands of the Civil War, both by water and rail (Bristol had had rail connections since the mid-1830s) resuscitated the local economy. The Keystone Forge, the Bristol Woolen Mills, and the other industrial firms in the town prospered as a result of massive government purchases. After 1865, however, the town's economy again declined as wartime markets evaporated; none of these businesses fully recovered from the post-war depression.

Joshua Peirce, a Bristol native who had spent several years in western Pennsylvania, returned to Bristol in 1868 and resolved to revive the town's depressed economy. In 1868 he purchased 49 acres of vacant land located on the outskirts of town between the Delaware Canal and the main line of the Pennsylvania Railroad, where he established the Livingstone Mills, a producer of felt products from wool. In addition to operating his own firm, Peirce encouraged other industrial enterprises to become established in the vicinity of his property. These firms, which included the Bristol Foundry, the Sherman Planing Mill (later known as the Sherman & Peirce Planing Mill

Bristol had the largest concentration along the Delaware Canal of mills and other industries, coal and lumber yards, and businesses directly related to the canal.

On the right is Lock 3, with a loaded Lehigh Coal and Navigation Company boat half hidden behind the mule about to enter the lock.

Below is the canal basin, and one of the mills on the north side of the basin.

and the Peirce & Williams Planing Mill) and the Bristol Rolling Mills, all contributed to the rejuvenation of Bristol's economy.

In 1876 Peirce and a number of other prominent local businessmen organized the Bristol Improvement Company, with the intention of offering "facilities to manufacturers desiring to locate here by erecting a building for their accommodation, thus encouraging the growth of manufacturing industries in the borough." As an additional incentive for relocation, all of the mill buildings were exempt from borough taxation for their first ten years of operation. Between 1876 and 1887 the BIC built five large mill complexes on portions of the 49-acre tract purchased by Peirce in 1868, which attracted textile manufacturers, leather goods makers, a wallpaper producer and the Standard Cast Iron Pipe & Foundry Company. The remaining portions of Peirce's original tract were apparently sold off and developed by others, principally as housing for the workers who labored in the mills and factories.

The most notable development was the arrival from Philadelphia of the woolen textile manufacturing firm Grundy Brothers & Campion, which leased the Bristol Worsted Mill, the first building erected by the BIC. The long-term success of this firm, which with over 1,000 workers eventually became Bristol's largest employer, made the name Grundy synonymous with the borough. The Grundy clock tower was a local landmark that could be seen for miles.[16]

Boatmen coming down the canal knew they were near the end of the trip when they saw the Grundy clock tower ahead of them. The 212-mile-long round trip from Mauch Chunk to Bristol and back took seven or eight days.

(*All photos these pages, NCM/D&L*)

Postcard view of the R & H Simon Silk Company's mill complex along the Bushkill Creek, Easton. Brothers Robert and Herman Simon started a silk mill in Union City, New Jersey, in the late 1870s. Lured to Easton by $35,000 in seed money put up by Henry Tombler and other prominent citizens, the brothers opened a silk ribbon mill in 1883. The firm expanded rapidly during the next two decades, weaving silk for dresses, neckties, and linings, as well as velvets, and silk plush for upholstery. By 1910 it had become the nation's largest producer of black silk ribbon. Simon Silk was rather unusual in that it handled all phases of silk production on one site, from the preparation of raw silk for weaving through dyeing, weaving, and finishing. This required a mill complex that had over 1 million square feet under roof, and employed as many as 2,000 people. Herman Simon died in 1913, and his wife Elizabeth ran the company until 1929. Various other firms wove silk in parts of the complex until 1981.

(NCM/D&L)

Chapter Nine

1880 to 1890
Steel, Cement—and Silk

*"I understood it all at once, approved of it, and made up my mind ...
to go into the manufacture of cement at once."*

José de Navarro

The 1880s saw two quantum leaps in industrial technology in the Corridor—the adoption of open-hearth steelmaking by Bethlehem Iron, and the introduction of the horizontal rotary cement kiln in Coplay. Bethlehem Iron also entered the armaments industry, designing and building a heavy forge capable of making armor plate for Navy ships. In addition, the Adelaide silk mill, a subsidiary of the Phoenix Silk Company based in Paterson, New Jersey, opened in 1881 with much fanfare in Allentown. This brought a new industry to the Corridor, one that would change the face—and gender—of industrial employment.

Bethlehem Iron Takes up Heavy Forging

In 1867, open-hearth technology was introduced into America by Abram S. Hewitt, the principal proprietor of the firm of Cooper and Hewitt, which operated a large iron and steel works at Trenton, New Jersey. The spread of open-hearth steel technology was not rapid; by 1880 less than ten percent of American steel was produced by this method. However, the growing use of open-hearth steel in applications such as bridge building, where a high tensile strength is required, soon spread to shipbuilding. By 1880, American shipbuilders and steelmakers had developed a body of useful experience in the use of open-hearth steel. During this time period, open-hearth steel became the preferred material for ordnance, armor plate, and propulsion machinery parts. Producing such items required heavy forging, but this technology was limited in the United States.

The availability and superior strength of open-hearth steel led to its adoption as specified material for the structural shapes and plates of the ships of the so-called "A.B.C.D. Squadron," four ships commissioned by the U.S. Navy in 1882, and built by the John Roach shipbuilding firm in Chester, Pennsylvania. The squadron was built as the direct result of several embarrassing incidents that showed the steep technical decline of the U.S. Navy since the Civil War. Its construction was fraught with technical and financial problems that uncovered an embarrassing dearth of technology and know-how in the American metallurgy industry.

Seeking solutions to the problem of a lack of domestic expertise, the Secretary of the Navy, William Chandler, and the Secretary of the Army, Robert Todd Lincoln, appointed a joint Army-Navy panel in 1883 to study the problems and propose solutions. The Gun Foundry Board made three proposals: first, that the Federal government supplement existing cannon-producing firms with additional tools and upgraded equipment so they could produce heavy ordnance; second, that the Federal government establish and own its own such facilities, while contracting with existing steel makers for forged and tempered open-hearth steel; or, third, that the government place large orders for heavy ordnance with American steelmakers, which would give them the necessary funds to invest in the needed technology.

The board then embarked on a tour of British and French cannon manufacturers to study their methods and dealings with their respective governments. They returned home convinced that the French method of buying semi-finished cannon forgings from private steelmakers and then converting them to finished artillery in government facilities was the most effective. Congress agreed, and designated the Watervliet Arsenal in New York and the Navy Yard in Washington, D.C., as sites for gun-finishing factories.

To provide semi-finished steel forgings, the War Department solicited bids from American steelmakers. This provided the opportunity for the Bethlehem Iron Company to construct the first and most important heavy-forging plant to be built in North America.

The Bethlehem Iron Company's decision to enter the military products field was the outcome of its search for new production items to compensate for the progressive erosion of its share of the market for steel rails. Although the company remained profitable, the long-term threat posed by the growing dominance of its Pittsburgh-area rivals was causing growing concern. John Fritz, who had long advocated diversifying the product line, had been strongly rebuffed by the directors when he suggested building a plant to roll structural members. His alternative proposal, to construct a large-capacity plate mill, was also rejected. Fritz then turned his attention toward winning the directors' approval of building a heavy-forging plant that was capable of producing military products.

While Bethlehem Iron was preparing to import the technology necessary to build America's first heavy-forging plant, the United States Navy took another large step in its campaign to create a viable modern fleet. During the spring of 1886, Congress passed a naval appropriations bill that authorized construction of two armored second-class battleships, one protected cruiser, one first-class torpedo boat, and the complete rebuilding and modernization of two Civil War-era monitors. Unlike the ships of the A.B.C.D. Squadron, the two second class battleships (U.S.S. Texas and U.S.S. Maine) would have both large-caliber guns (12-inch and 10-inch respectively) and heavy armor plate. On August 21, 1886, the Navy solicited initial bids for 1,310 tons of semi-finished gun forgings and 4,500 tons of steel armor plate. After a period of intensive debate in the Navy Department, a second and definitive solicitation for the ordnance forgings and armor was issued on February 12, 1887.

The solicitation allowed interested firms to submit bids on either gun forgings or armor plate,

John Fritz installed open-hearth furnace steelmaking in Bethlehem in 1888 in order to make steel suitable for forging into both guns and armor for the US Navy.

This 1895 photograph from Fritz's personal collection shows workers pouring a steel armor-plate ingot in Open Hearth No. 1. Open-hearth steelmaking was much slower than Bessemer production, but the qualities desired in the final product could be controlled in steel made using this process.

(All photos NCM/D&L)

Left, the 125-ton steam-powered drop hammer forging a large piece of 17-inch-thick armor plate in the early 1890s. Designed by John Fritz, it was the largest forging device in the world, but the noise and impacts while it was operating knocked machine tools in the plant out of calibration, and shook nearby houses and their residents. The hammer was too big for its striking force to be controlled, and it was taken out of service in 1894.

Below, the 125-ton hammer stands in the shadows in the background as its replacement, the 14,000-ton steam-hydraulic forging press, forges its first piece of armor plate in 1894. The largest open-die forging machine ever constructed, it was used to make large-caliber cannon as well.

It remained in operation through both world wars and was scrapped in 1954.

or make a combined offer to manufacture both products. It was implied that preference would be given to companies that submitted combined bids, but of the four American steel companies that offered proposals, only Bethlehem Iron did so. After lowering its bid on gun forgings to meet a lower offer from Cambria, Bethlehem successfully secured both the forgings and armor contracts on June 28, 1887.

In order to fulfill the contracts, between 1887 and 1892 John Fritz introduced open-hearth steelmaking in Bethlehem, and then designed and completed America's first heavy-forging plant.

Open-hearth steelmaking was a slower process than Bessemer; however, it allowed the company to carefully control the qualities of its steel. The massive expansion of the plant and its workforce was a costly gamble, but company directors Robert Sayre, E.P. Wilbur, and Joseph Wharton wisely backed Fritz, thus ensuring the survival of the company.*

Fritz designed the forging plant, an enormous expansion in the capacity of the Bethlehem works, with the able assistance of Russell Wheeler Davenport. The building containing the open-hearth furnaces, forging presses, fluid compression apparatus, and plate mill was 1,155 feet long by 111 feet wide, and included a 125-ton steam hammer, open-hearth furnaces with the capacity to cast 100-ton ingots, and cranes with up to 150-ton capacity.

By the autumn of 1890, Bethlehem Iron was delivering gun forgings to the U.S. Navy and turned its energies to completing the facilities that were necessary to produce armor plate. John Fritz began designing the 125-ton steam hammer that was to symbolize both the company and the country in the next decade.[1]

Cement Expands in the Valley

After a full decade of being the only portland cement maker in the United States, Saylor Cement was faced with new competition from only a few miles away. In 1883, the American Improved Cements Company was founded near Egypt, Pennsylvania, by John Trinkle and Robert Lesley, who were importers of European portland cement.

His established business notwithstanding, Lesley thought that the market for natural cement was stronger, and he walked the length of the Jacksonburg formation from Martins Creek westward, testing samples of rock and looking for a site for a cement mill. (Much to his later regret, he rejected Northampton County's high lime content stone that was soon bought up by portland cement makers such as Alpha, Bath, and Dexter.) He found the stone he was looking for in the village of Egypt, and built a mill there that produced both natural and portland cements.

After visiting the Bradley Pulverizer Company's factory in Wymouth, Massachusetts, and being impressed with the efficiency of that company's equipment (at that time being used to make fertilizer), Lesley installed one of their Griffin mills in his plant in Egypt, the first time one was used for cement making. A few years later, Thomas Whittaker, founder of the first portland cement mill in New Jersey, spent two days and two sleepless nights on a hill overlooking the American mill, listening to the "booming noise of the constantly revolving Griffin Mill." Lesley recounted in his history of the American cement industry, "When at the end of that time, the mills were still rolling along merrily and he had heard no stoppage (a common problem in rock crushing at the time), he went home and put in his order for similar mills."[2]

Bradley's Griffin mills became the standard for crushing and grinding equipment in the cement industry. In a far-sighted move, the company relocated to Allentown in 1886 to serve the

---

* Most of the many other iron furnaces in the Corridor, all of which started up again some years after the Panic of 1873—some of them as reorganized companies—remained in production through the end of the century, but none of them moved beyond the production of pig iron. The last of the merchant pig furnaces, as they were called, closed in the 1920s. Thomas Iron stopped in 1924, and the Crane was completely shut down by 1930.

A Bradley pulverizer on the factory floor of the firm's plant in Allentown (right), and at work in an unidentified cement plant in a photograph dated July 23, 1922.

(Photographs courtesy of Bradley Pulverizer, Inc.)

cement industry. Now a multinational corporation that does business around the world, Bradley Pulverizer's headquarters remain in Allentown.

Big Kiln Keep on Turnin'

An even more important development in cement production took place in Coplay only three years later—the introduction of the horizontal rotary cement kiln. José de Navarro was a Spanish-born New York City entrepreneur and industrialist who had become interested in cement while constructing the Metropolitan Elevated Railway and the ill-fated Navarro Flats apartment building, both in Manhattan. Learning of the simultaneous developments of horizontal rotary cement kilns by English engineer Frederick Ransome and American Henry Mathey, he set out to purchase rights to both. With the American rights in hand by 1885, Navarro and his sons purchased a plant in Rondout, New York, that had been making the Rosendale type of cement and installed a 24-foot-long, 12-foot-diameter horizontal kiln there. Despite the fact that their kiln could produce far more cement than conventional upright stationary kilns, the Navarros found that it was not cost-effective and sold the plant. Navarro then sent his younger son to Britain with George Collingsworth, the expert behind the Ingersoll Rock Drill's success (another of Navarro's successful enterprises) to acquire the American rights to Ransome's patents.

The first successful American horizontal rotary cement kiln, on site in Coplay in the "Valley" mill of Atlas Cement shortly before the building was demolished in the early 1930s. By then, the kiln that made all vertical cement kilns immediately obsolete had been operating for over forty years.

Coplay Cement letterhead, circa 1900, shows Mill B, with the Schoefer kilns in the foreground, and the original Mill A slightly upstream. An Ironton Railroad train rides the tracks that curved past the plant in the extreme lower right of the image.

In 1888, Navarro leased 17½ acres of land on the Lehigh River in Coplay, just north of the Saylor Cement works. He incorporated the Keystone Cement Company (which had no relationship to the firm begun in Bath, Pennsylvania, in 1928, which is still in operation) and began construction of two rotary kilns in February, 1889. The first kiln was put into operation on November 8 of that year, and the 125 barrels of cement produced that day were the first portland cement made by the rotary process in the United States.

The rotary kiln was capable of producing more cement in ten to twelve hours than a stationary upright kiln could produce in a week. Nevertheless, the American cement industry, then concentrated in the Lehigh Valley, was so resistant to the rotary kiln that the Saylor Cement Company

José de Navarro, 1823–1908

José Francisco de Navarro Arzac was born into an aristocratic though poor family in San Sebastian, in the Basque region of northwest Spain. Admitted to the Spanish Royal Naval Academy at the early age of 12, Navarro was too young to be commissioned an officer when he graduated at 15. So he sailed alone to Cuba to his mother's brother, a sugarcane grower, but confounded his uncle by preferring to work with the mechanics who kept the sugar-processing machinery functioning, rather than as a manager. After a year of this work, Navarro's lifelong grasp of and fascination with mechanical technology was cemented.

The next year Navarro attended the University of Havana with the goal of learning enough English to attend Stephen Van Rensselaer's Troy Institute. After arriving in Philadelphia in 1841, he found his English skills inadequate, then fell ill with scarlet fever. The Chase family of Baltimore, who had trade connections to Cuba, took him into their family and business for two years. They taught him the shipping and warehousing business, and gave him time to study the Troy Institute's full engineering course books, a gift from Van Rensselaer.

Armed with this experience and knowledge, Navarro returned to Cuba; at age 20, he ironed out the tangled warehouse and wharf logistics in Cárdenas, the major sugar-shipping port. In the process he bought a significant amount of land at a low price and partnered with a major sugar grower, though he had to wait until he was 21 to make the partnership legal.

His sugar-export business had eight very successful years, but Cuba's instability worried Navarro, so he sold his half of the business to his partner. Then disaster struck: a fire destroyed most of Cárdenas, including all the ships, wharves, and warehouses they owned, and the entire sugar crop they were preparing to send off. With no insurance, it took Navarro a year to make good for all the losses. Armed with about $10,000 instead of the half million he had had the year before, he arrived in New York in 1854.

Navarro quickly rose to the top of the mercantile class in New York, and went into a shipping partnership with "Commodore" Cornelius Garrison. For the next 30 years, he moved into many enterprises: warehousing; manufacturing an innovative water meter that he invented; perfecting and manufacturing the Ingersoll Rock Drill; becoming a founding partner in both Equitable Life Insurance and the Edison Electric Light companies; and building the Sixth Avenue Elevated Railroad, the first such successful system in the U.S.

No obstacle or crisis ever seemed to daunt or discourage him. Corrupt New York politics and national financial panics cost him millions in lost assets over the years, none worse than the failure of—and his personal liabilities in—the massive Central Park Apartments. Also known as the Navarro Flats, this was a huge complex of cooperative luxury apartments, the first in New York to be fully electrified and with hot-water plumbing. It became a nightmare of lawsuits, broken contracts, and defaulted mortgages, leaving Navarro deeply in debt. Forced to sell many assets and to turn to his independently wealthy wife for assistance, Navarro then began his foray into cement.

Navarro married Ellen Amelia Dykers, the daughter of John Hudson Dykers, one of the founders of the New York and Harlem Railroad, in 1857. Marrying into the most prominent and wealthy Catholic family in New York secured his place in the city's society, which might have been difficult during the anti-immigrant, anti-Catholic sentiment prevalent at the time. Only two sons survived into adulthood, Alfonso and Antonio; both went into the cement business with Navarro, but only Alfonso stayed in it.

Navarro died in New York in February 1909, and is buried in Calvary Cemetery, Queens. Despite his deep involvement in many important nineteenth-century developments and innovations, his grave and three stained-glass windows in Holy Cross Catholic Church in Rumson, New Jersey, where the Navarros had a summer home, are his only memorials—as well as every rotary cement kiln in the U.S.[3]

erected ten Schoefer kilns—as it turned out, the final innovation in stationary vertical kilns—a quarter of a mile away from Navarro's cement works in 1893. This was a full four years after Navarro's horizontal kiln began turning out vastly more cement, cheaper, and in much less time. These kilns, the lower parts of which still stand in Coplay, were used for only about ten years.

Despite being made from the world's best natural deposits of cement rock for making portland cement, domestic cement was still considered inferior to imported British and German products. Navarro became determined to make a portland cement that was at least 50 percent better than any made in America, so after a period of experimentation, he subjected his cement to the same tests used by the Stettin Company of Germany, then the makers of the world's best portland cement. He further quelled doubters by asking for the U.S. Government's evaluation. The federal authorities not only validated Navarro's claims, but placed large orders for his product.

José de Navarro

Keystone Cement sold cement as fast as it could make it for the next two years, until Navarro's continuing legal and financial problems from the Navarro Flats project forced him into bankruptcy. Rescued by his independently wealthy wife and sons, at the end of 1891 Navarro sold a controlling interest in his plant, by then renamed the Atlas Cement Company, to a group of investors headed by John Rogers Maxwell, president of the Central Railroad of New Jersey.

Relieved of the burden of overseeing the company's finances, Navarro turned his attention to ways to improve their product. In his business memoir, he takes credit for about a dozen improvements, including two that he and others in the industry regarded as breakthroughs. The first was the addition of a cooling cylinder; the second was adoption of pulverized coal as a substitute for petroleum as the fuel for heating the kiln. An even earlier "improvement" was the most important. In the first days of Keystone's operations, Navarro retained the services of a French chemist named Giron, who was working on ways to regulate the rates at which concrete would set by adding gypsum. Giron's work allowed Keystone to inform its customers in advance how long it would take for their concrete to set.[4]

In his history of the cement industry, Robert Lesley describes Navarro's role:

The successful manufacture of portland cement with pulverized coal under the rotary kiln process was so marked as to practically terminate competition between the rival types of kilns, thus turning the entire American industry over to the rotary kiln which is now in use in every mill in the country.[5]

With many of the technical problems thus solved, and demand for the product rising exponentially, the cement industry was poised for a rapid expansion in the decade of the 1890s.

Silk Weaves Its Way into the Corridor ...

On November 17, 1881, the Adelaide Silk Mill, a venture of the Phoenix Silk Company of Paterson, New Jersey, opened with great fanfare in Allentown. The city's mayor expressed the hope that it would be "the nucleus of a great and extensive enterprise in the Lehigh Valley."[6] That hope was fulfilled, though not immediately. It was not until the first decade of the twentieth century that silk manufacture became a major part of the Corridor's economy, propelling both Pennsylvania and the United States into the dominant position in the global industry.

In 1881, the city fathers of Allentown had to work hard to attract Phoenix Silk to the city. They were driven by the protracted economic depression that had gripped the city after the Panic of 1873, which had closed many of the Lehigh Valley's iron furnaces for a period of some years.

Paterson, on the other hand, was the bustling "Silk City," home of more than two dozen silk mills and related industries. Nevertheless, the industry's leaders were also facing challenges. Efficient new machinery for taking silk from its raw state to beautiful fabric made it possible to phase out many of the skilled, mostly male, artisan silk workers who had dominated the labor force in Paterson since the 1840s. Needless to say, they were fiercely resisting the trend to lower-skilled (and lower-paid) workers. These men, mostly of British and French birth or ancestry, and their bosses were also challenged by the influx of experienced silk workers from Eastern Europe, many of whom were not only skilled but tended to hold disturbing socialist ideas about workers' rights and fair pay.

All this led the Paterson silk makers to look beyond their city for new locations to make silk. They were hoping to find cheap energy, a docile workforce accustomed to factory jobs, and easy transportation to both Philadelphia and New York.

Allentown's city fathers promised all three, as well as money to help Phoenix build a mill on the banks of the Jordan Creek. It was an uphill battle in a bad economy for the city's business leaders to raise the funds privately, but they succeeded. The Adelaide Mill, named in honor of the wife of Phoenix Silk owner Albert Tilt, opened first as a winding and spinning mill—all lower-skill jobs—to supply silk yarn for the looms, which remained in Paterson, but it soon became a weaving mill that operated into the late 1950s. (The building, which still stands along the Jordan between Hamilton and Linden streets, was expanded several times, and was briefly the largest masonry building in the United States.)

The 1881 Adelaide Mill along the Jordan Creek in Allentown. This was the first of many silk mills in the Lehigh Valley.

The many components of the silk industry came to dominate the local economy in the Corridor for decades.

(Courtesy of of Kelly Ann Butterbaugh)

In 1888, three silk mills opened in Bethlehem, one in each of the boroughs of Bethlehem, South Bethlehem, and West Bethlehem. Fichter and Martin Ribbon Company opened the business that soon became the mammoth Bethlehem Silk Company mills on West Goepp Street; Lipps and Sutton began weaving operations at Seneca and Clewell streets in South Bethlehem, and John D. Cutter and Co. opened a mill in West Bethlehem along the Lehigh Canal that became one of the Saquoit Company's weaving mills.

The year before, tiny Weatherly landed a mill owned by Paterson's Read & Lovatt Silk Manufacturing Company over several much larger competitors by putting up $35,000 of the $50,000 cost for a new mill building and a rail-line connection to the Lehigh Valley Railroad. Weatherly got its money's worth; within a year, Read & Lovatt employed 225 townspeople (out of a population of 2,500), each earning $1.50 per week. Rather than producing woven silk, the company was a "throwster"—performing the low-skill processes of preparing raw silk for dyeing, weaving, or knitting.

These mills established a geographic pattern for the industry, which persisted as long as Pennsylvania's silk industry did. In general, throwing and winding mills were found in the northern counties, such as Luzerne, Carbon, Lackawanna, and Schuylkill, while weaving mills clustered in Northampton, Lehigh, Berks, and Lancaster counties. Knitting mills, making things such as hosiery and gloves, centered in Bucks and Montgomery counties. The reason for this was labor costs, and keeping the labor peace. Weavers and warpers were high-skilled workers who commanded higher wages than the girls—often children as young as 10—who could be taught to mind machines that spun or wound silk yarn. Mill owners looked to forestall labor troubles by keeping workers with such disparate earnings far apart.

Silk was the first industry to develop in Pennsylvania because of the Commonwealth's human resources, rather than merely its natural ones. Especially in the coal regions, but also in the Lehigh Valley, Paterson's silk magnates found a ready supply of docile, fairly well-educated female labor. Work in silk mills—relatively safe, relatively comfortable, and relatively non-strenuous—provided the first industrial work opportunities for females. Mill work offered better pay, a more respectable environment, and somewhat more independence for women and girls than the domestic-service jobs that were often the only alternative, especially for newly arrived immigrants. From a handful of mills spread between Philadelphia and Scranton, Easton and Reading, silk began to flow in an increasing stream—one that would lead eastern Pennsylvania to dominate the entire industry in the first decades of the twentieth century.[7]

Industrial Tourism

The Switchback and the "Switzerland of America"

Tourists, including writers, artists, and influential, wealthy people had been coming to Mauch Chunk since the 1820s. They were attracted to the beauty, and fascinated with new technology and especially the Gravity Railroad. By 1830 the Mansion House Hotel in the town had 50 rooms, and was able to accommodate business travelers as well as increasing numbers of tourists, many of whom came to leave the stifling summer heat of Philadelphia and enjoy the pleasant coolness of the mountains. Exaggerated descriptions and fantastical images of the area were published in newspapers and perodicals. This engraving by Harry Fenn, published in Volume I of the subscription publication *Picturesque America,* was titled "A Mauch-Chunk Highway."

The Lehigh Coal & Navigation Company promoted the public interest, and allowed thrill-seeking passengers to ride the Gravity Railroad from its early years. Mauch Chunk became one of the few places in America that successfully combined tourism with an industrial setting. Historian John F. Sears concluded that its popularity resulted from the fact that at Mauch Chunk, industrial activities blended with natural scenery to create an atmosphere that confirmed the faith of Americans in technology, progress, and wealth.[8]

In 1872 the Nesquehoning Tunnel opened, connecting the Panther Valley mines to the Coalport loading docks upstream from Mauch Chunk and rendering the Switchback obsolete as a coal hauler. Residents of Mauch Chunk, particularly the owners of hotels and restaurants, feared that LC&N would tear down their chief tourist attraction; instead, the company decided to improve it. The cars that carried passengers were made more comfortable, and an elegant new pavilion on top of Mount Pisgah became a destination for dancing and refreshments in the cool, clear mountain air. These improvements helped propel the Switchback's ridership to nearly 12,000 passengers in June and July of 1873. Then, the financial difficulties that pummeled the national economy during the years following the Panic of 1873, coupled with a violent railroad strike in 1877, kept tourists away and even bankrupted the owner of the Mansion House.

Prosperity returned in the 1880s and 1890s, as thousands of visitors came for the cool breezes and beautiful mountain vistas of the "Switzerland of America." In 1885, the *Carbon Advocate* estimated that 100,000 visitors had come to ride the Switchback. Though escaping the summer city heat lured many visitors, the real peak came in the autumn: five or six thousand visitors poured off Lehigh Valley and New Jersey Central passenger cars and headed for the 18-mile thrill ride virtually every day in October, 1893. By the close of the nineteenth century, the Switchback was carrying well over 100,000 passengers a year. These were not day trippers—they crowded the hotels and restaurants, bought souvenirs and sent post cards, making Mauch Chunk one of America's greatest tourist destinations.

Competition came early in the twentieth century, when an electric trolley line from Lehighton ran up adjacent Flag Staff Mountain, which boasted views equal to the Switchback's sights from Mount Pisgah. A park with a dancing pavilion (the Dorsey Brothers from Lansford honed their swing sound there) and picnic grounds soon followed. People and organizations moved their excursions there.

The Switchback was further hurt when the railroads had to eliminate much of their passenger service in order to haul coal, troops, and essential war goods during World War I, followed by the advent of private automobiles in the 1920s. Profits shrank, then moved into the negative column, and the Great Depression was the final blow.

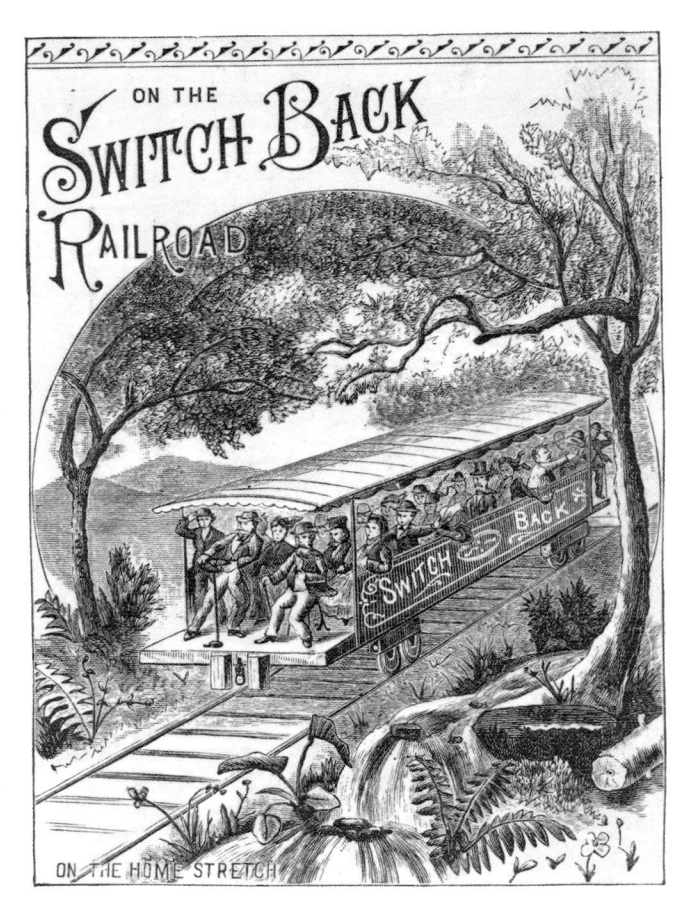

The last car rode over the rails from Summit Hill to Mauch Chunk on October 29, 1933, and the railroad was scrapped four years later.[9]

Chapter Ten

1890 to 1900
Armor, Immigration, and Cement

"Armor plate was one of the things the Government must have, and as iron was useless in front of modern steel shot and shells, steel must be the material ... "

John Fritz, The Autobiography of John Fritz

In the 1890s, the industrial fame and reputation of several of the Corridor's major industries began to spread not only across the United States, but to Europe and beyond. The Bethlehem Iron Company, guided by the engineering genius of John Fritz, moved into the international arena as its armor plate and large guns for the United States Navy attracted the attention of foreign naval forces. The cement kilns of the Lehigh District turned out 60 percent of the millions of barrels of portland cement being produced in the United States, and largely ended the importation of foreign cement. And the innovative Lehigh Zinc Company was absorbed by its leading rival, New Jersey Zinc, which began its ascent to national dominance of its industry.

Armor Plate and the 125-ton Steam Hammer

By the autumn of 1890, Bethlehem Iron was successfully delivering gun forgings to the U.S. Navy and was thus able to devote its energies to the completion of the facilities that would be necessary to produce armor plate.

During the early 1880s a vigorous debate took place among the world's naval officers concerning the relative merits of the accepted compound armor, in which a thin hard steel plate was welded to a thicker and more elastic backing of wrought iron, versus the newly developed homogeneous steel armor that was produced by the French manufacturer Schneider et Cie. Test bombardments of compound and homogeneous plates were undertaken by the United States Navy and other maritime forces to compare the merits of each process. The results demonstrated that compound armor was prone to both cracking and penetration after repeated hits, while the Schneider homogeneous steel armor could be penetrated but not shattered by prolonged bombardment. It was also lighter in weight. Directors of the Bethlehem Iron Company recognized the advantages of the Schneider armor, and by December of 1886 formed a contractual alliance with Schneider in order to obtain U.S. production rights. In the Schneider process, specially prepared plates made with open-hearth steel were forged using a heavy 100-ton steam drop hammer. Bethlehem's foresight in acquiring the production rights was rewarded when the U.S. Navy decided, after examining the armor-plate test results, to specify homogeneous steel armor for its warships that were under construction.

During the 1890s Bethlehem Iron Company became the leading manufacturer of armor plate in the U.S.

On the right, a worker is removing scale from a massive piece of armor plate.

Below, a steel ingot with a mandrel inserted in the center is being shaped by a 7,500-ton-force forging press, 1894. Forging cannon and armor plate began at Bethlehem Iron in the 1880s and continued through much of the twentieth century.

(NCM/D&L)

John Fritz realized that if the striking power of the forging hammer could be upgraded, a superior product would result. He designed a monstrous 125-ton steam hammer, the largest forging device of its type ever to be constructed. This gigantic device was installed at the forge shop of the Bethlehem plant and soon became famous. A wooden model of it became the centerpiece of Bethlehem's exhibit at the 1892–1893 Columbian Exposition in Chicago. The unprecedented size and complicated mechanics of the great hammer greatly delayed its completion. Begun in 1889, it did not become operational until June 1891. After so much effort, its operational life was short. Within three years of its startup it was permanently idled. Several reasons are cited: Eugene G. Grace, who later became chairman of the Bethlehem Steel Corporation, wrote in his article "Making Ordnance at Bethlehem" that the shock of the hammer's blows continually moved the large gun-boring and -turning machines located in a nearby machine shop out of alignment. (Neighbors in surrounding South Bethlehem complained that their houses shook as well.) A more cogent reason for its demise is that Bethlehem installed a powerful hydraulic forging press in 1894. This was better able to deform and shape deeper cross sections of large armor-plate ingots. The great hammer was scrapped sometime between 1901 and 1903.

Bethlehem Iron Company's 14,000-ton press had two hydraulic forging cylinders, each 50 inches in diameter. The cylinders gave the press a forging stroke of 816 inches. The heads of the press were made in two pieces, bolted together by 18 steel bolts, each 6 inches in diameter. Before the heads were bolted, the individual bolts were heated to a red heat. After they had cooled, the bolts were screwed home with a pressure of 20,000 pounds per square inch.

The four great column bolts of the press were each 40 feet long and 26 inches in diameter. The entire press was over 47 feet in height and could handle ingots up to 14 feet by 14 feet. The dies, forging blocks, and other tools of the press were manipulated by hydraulic power. The press was served by two hydraulic cranes with a capacity of 65 tons each. The hydraulic pressure was supplied by a 15,000 horsepower three-cylinder vertical steam pumping engine designed by Fritz in consultation with engineer E.D. Leavitt of the I.P. Morris Company of Philadelphia.

Each of the engine's three simple cylinders were 54 inches in diameter and 90 inches high. The engine had a 50-inch stroke and operated at a steam pressure of 150 psi. Steam was supplied by 32 Leavitt boilers, which were housed in a separate building. The engine employed Stephenson valve gear and operated at the low speed of 80 revolutions per minute. The pumps that delivered hydraulic pressure to the press were attached horizontally to the back of the engine and received their motion from the cylinders through bell cranks with no power being transmitted to the shafts except to and from the flywheels. The pumps supplied a hydraulic pressure of 7,000 pounds directly to each of the cylinders. (NCM/D&L)

The enormous 14,000-ton-force hydraulic forging press that replaced the hammer was used almost exclusively for producing armor plate and the largest ordnance and commercial forgings. Designed by John Fritz, it was his last major work for the Bethlehem Iron Company and was constructed to be of great strength and durability. Its dimensions were virtually unprecedented in the United States.

Despite the mechanical genius of John Fritz and the metallurgical knowledge of Russell W. Davenport, the Bethlehem Iron Company experienced great difficulties starting production of armor plate. Although Fritz and Davenport built on the best contemporary European forging technology to produce a plant of unequaled size and capacity, the problems of forging armor plate were not easily solved. By 1890, it became clear to the U.S. Navy that Bethlehem would not be able to meet its contract deadlines for delivering armor plate, a delay that greatly hindered the completion of such major warships as the first modern American battleships, U.S.S. *Maine* and U.S.S. *Texas*.

The government's planned expansion of the American fleet necessitated construction of an additional armor-plate shop; in November 1890, Secretary of the Navy Benjamin Franklin Tracy signed a contract with Carnegie Phipps Foundry of Pittsburgh for 6,000 tons of armor plate. Significantly, the contract specified that the plate could be manufactured from simple steel or

Production of steel rails was becoming increasingly less important in Bethlehem as western Pennsylvania steelmakers were increasing their share of that business. Fritz and his team, understanding that they had to find new products to remain profitable, moved Bethlehem Iron aggressively and very successfully into the armaments business. They made armor for the navies of other nations in addition to the U.S. Navy. On the right are pieces of armor plate for the Russian navy.

A testing ground for cannon and armor plate was set up at an abandoned limestone quarry at Redington in 1887. Representatives of foreign countries looking for arms and armor would be brought here to see on-site demonstrations of what Bethlehem was capable of providing.

The ability of Bethlehem Iron to forge large objects brought a variety of orders. Bethlehem came to dominate the market for large forgings for the fast-growing electric power industry. A completed field ring for the Niagara power plant is below right, in 1900. Below is the 45-ft-long, 89,320-lb axle for the Ferris wheel, which was the biggest sensation at the Chicago World's Fair in 1893. It was the largest forging in the world to date. Bethlehem made many other parts for the wheel.

(All photos this page, NCM/D&L)

nickel steel. But the following year, American industrialist Hayward Augustus Harvey, who had invented a tool-steel hardening process, began construction of a face-hardening furnace in the armor-finishing No. 3 Machine Shop of the Bethlehem Iron Company. The Navy had put much financial support behind Harvey's process, and when the furnace produced the first commercial face-hardened armor plate in America on July 30, 1892, at Bethlehem, the Navy conducted tests of samples at Bethlehem's Redington proving ground. The plate successfully resisted the impact of five 250-pound armor-piercing projectiles, fired from an 8-inch gun. All of the projectiles broke apart and penetrated to a depth of only three inches. The combination of nickel steel and face hardening produced what became the basis for almost all of the armor plate produced by both Bethlehem and Carnegie.

Winning Wars

Despite competition from the Carnegie Phipps armor mill, Bethlehem Iron Company continued to be profitable. During July of 1892 and March of 1893 Bethlehem received large orders from the Navy for gun forgings and armor plate. To further increase their return on the investment they had made in the forging plant, the company's directors entered the international ordnance and armor market. In 1894 their efforts were rewarded when they received a contract for 1,200 tons of armor plate from the Imperial Russian Navy. By 1895 the forging and treatment plant of the Bethlehem plant had become internationally renowned as a leader in the production of ordnance forgings and armor plate. Lieutenant Colonel W. Hope, V.C., a British ordnance expert, visited the Bethlehem plant as a part of his worldwide tour of ordnance works and declared, "I consider the Bethlehem Gun Plant to be superior to any gun plant in the world."

During the remainder of the 1890s Bethlehem Iron continued to broaden the range of its forged products. The company dominated the market for large forgings for the electric power industry and it produced the field rings and rotor shafts for the large hydroelectric plants that were being built near Niagara Falls. The company's technological expertise was also demonstrated by the forging of important components for the great Ferris wheel of the 1892–1893 Chicago World's Fair—the axle of the wheel was the largest steel forging to be manufactured up to that date. The company also provided crank shafts and propeller shafts for the engines of many of America's passenger liners and merchant ships.

Bethlehem Iron also manufactured finished ordnance and gun forgings. Almost all of Bethlehem's orders for finished heavy ordnance were placed by foreign armed forces, while the United States continued its established practice of purchasing semi-finished forgings to be machined and finished at the Washington Naval Gun Factory and the Watervliet Arsenal. By 1900 both the U.S. Army and Navy had begun to purchase finished ordnance from the Bethlehem Iron Company, necessary because of the rapid increase in the number of coastal-defense batteries and large warships. Bethlehem Iron Company's No. 2 Machine Shop assumed the role of a major gun-manufacturing facility and by 1900 it had become one of the largest and best-equipped machine shops in the world.

A train of coastal defense guns leaving Bethlehem on Lehigh Valley Railroad flat cars to be finished at Watervliet Arsenal in New York State. Late 1890s.

Bethlehem also made Buffington Crozier disappearing carriages (left). These could be dropped down below the fortifications of the coastal defense stations, hiding the large guns that were mounted on them.

(Both photos: Fritz collection, NCM/D&L)

The success of American warships during the short and victorious Spanish-American War served as a great advertisement for the Bethlehem Iron Company, since many of the vessels contained armor, ordnance, propulsion machinery parts, and propeller shafts that had been manufactured by the company. The high regard in which Bethlehem was held by the United States government is evident in the invitation from President William McKinley to Robert H. Sayre to take a prominent place on the official reviewing stand as the victorious United States Atlantic Squadron passed in a stately procession up the Hudson on August 20, 1898.[1]

The Third Wave of Immigration

By 1899, Bethlehem Iron had over 7,000 employees, and 10,000 people lived in the borough of South Bethlehem. One-quarter of them had been born abroad, a higher percentage than the 16 percent of foreign-born residents in the entire state of Pennsylvania. Those 2,600 immigrants represented 64 ethnic groups.[2] The majority of them were Roman or Orthodox Catholics.

A similar picture was developing in the coal regions, particularly in Luzerne County. As the United States' economy began to shake off the tight-money shackles caused by the Panic of 1873, immigrants from southern and eastern Europe learned about the opportunities available in Pennsylvania. Fleeing war, conscription, and poverty (and often lured by an overly rosy picture of life and work in America painted by agents hired by the railroads and industries like steel and cement), Poles, Italians, Lithuanians, Slovaks, Ukrainians and Ruthenians, and many thousands of eastern European and Russian Jews began arriving after 1881.

What historians and anthropologists refer to as the "Third Wave" of U.S. immigration was fueled by several factors in addition to active recruitment. Improvements in the speed and size of

Political cartoonist William Allen Rogers was not always sympathetic to the immigrants flocking to America during his tenure (1877–1902) as the chief illustrator of Harper's Weekly, *then the most widely read magazine in the country. But this lithograph of exhausted, apprehensive immigrants, clearly depicted as southern and eastern Europeans, seems to express empathy for their harrowing journey and the uncertainty of their futures in the coal patch on the nearby hill. The article that accompanies this image in the July 28, 1888, issue decries the common though illegal practice at the time of "importing" workers from foreign countries to work in mines, quarries, and other low-skilled but dangerous jobs by promising them steady paid work. The article identifies the standing man with the umbrella as the company's agent, who collected these people when they arrived in New York and has brought them to his—and now their—employer. They will certainly find that the work is harder and the pay much less than they contracted for. The article blames this practice for the steady erosion of workers' wages in Pennsylvania during the 1870s and 1880s, and condemns it as not only "white slavery" but damaging to the economic progress of the state and the nation.*

transoceanic ships made it cheap enough for even poorer people to emigrate. Nearly a decade of intermittent wars of rebellion against the Ottoman Empire caused many people in Bulgaria, Romania, and the Balkans to flee their homelands, while pogroms against Russian Jews unfairly accused of being responsible for the assassination of Czar Alexander II drove millions of Jews from Russia and Russian-controlled Poland. Most of these southern and eastern European immigrants arrived via New York through the immigration portal of Castle Garden in lower Manhattan and, after 1892, Ellis Island, and found factory jobs in cities in the Northeast and Midwest.

The Third Wave, between 1880 and 1914, brought over 20 million European immigrants to the United States, an average of 650,000 a year, at a time when the U.S. had 75 million residents.

Irish laborers in the Weissport area, believed to be working on either the Lehigh Valley or the Lehigh & Susquehanna Railroad. A handcar is visible behind them.

(NCM/D&L)

Irish immigration had leveled off following the Civil War years, and later Irish immigrants faced considerably less hostility than those who had arrived during the 1850s. Part of the reason was that thousands of young Irish men (admittedly, many of them conscripted) had served in the Union Army, which led to quicker assimilation and greater acceptance by Americans. Assimilation was speeded by Irish neighborhoods that were large enough to develop economic and political clout by the late nineteenth century in the industrial cities of the North and Midwest.

By the late nineteenth century, many Irish and Irish-Americans were gaining some measure of upward mobility, but many were still shunted into low-paying, low-status jobs. Now they were competing with millions of new immigrants from Italy (the largest single group of immigrants between 1880 and 1920), Russia, Austro-Hungary, and the Balkans. In a pattern that repeats itself to this day, many of these newcomers found that the only work available to them was the physically demanding, poorly paid jobs that few native-born Americans were willing to do.

Like the Irish a generation earlier, the newcomers, mostly illiterate peasants forced off their land, were often the targets of bigotry and discrimination. In 1899, *The Allentown Daily Leader* described Hungarians in the Catasauqua area as "the Huns … a bad and reckless lot, every one of them carrying deadly weapons, which are used on the slightest provocation."[3] Three years earlier, the same paper noted approvingly that two of the borough's biggest industries, the Bryden Horse Shoe Works and the Davies & Thomas Foundry, "can boast of never employing a Hungarian workman."[4] Ironically, it was the Hungarian laborers at the Crane Iron Works, toiling thirteen hours a day on the casting floor and other hot, dirty jobs for ten cents an hour, who were steady, reliable workers with no interest in supporting their fellows in the Bethlehem Steel strike of 1910.[5]

Many Italian immigrants had a skill such as tailoring, barbering, stone cutting, masonry, or textile dyeing and weaving that allowed them to find work.[6] But "imported" Italian laborers in the anthracite regions were not only exploited in the collieries but vilified in the press and even by the government. In 1884, the Pennsylvania Commissioner of Industrial Relations disparaged Italian coal workers: "The illiteracy, turpitude, and degraded habit of this class of immigrants, innate and lasting as they are, stamp them as a most undesirable set." The fact that many Italian miners actively supported the labor movement and the push for mine safety set them at odds with powerful interests in Pennsylvania until the collapse of the anthracite industry.[7]

European immigration was slowed first by World War I and then by numerical quotas in the 1920s.

Overreach on the Anthracite Railroads

At least one-third of the anthracite mined in Pennsylvania between 1880 and 1900 moved to market on transportation routes that went from the Wyoming and Lehigh coalfields through the Lehigh Valley. These routes were, in order of importance, the Lehigh Valley Railroad, the Lehigh and Susquehanna–Central Railroad of New Jersey system, and the Lehigh Navigation. The bulk of the coal shipped by the Lehigh Valley Railroad came from mines located in the Lehigh field near or at the city of Hazleton, though some came from the Wyoming Valley. The anthracite shipped on the Lehigh and Susquehanna–CNJ Railroad system came mostly from the Wyoming Valley, with a smaller percentage coming out of the Panther Valley mines. All of the anthracite transported on the Lehigh Navigation came from Panther Valley mines.

Although the 1870s and 1880s witnessed the continuing expansion of Pennsylvania's anthracite production, its markets grew at a much slower rate. As a result, competition among the various anthracite-mining and -transportation companies grew ever more intense. The inevitable result was a fall in profits. Under the leadership of Franklin B. Gowen (1836–1889), the Philadelphia and Reading Railroad, which largely controlled the Schuylkill field and was the largest single producer and shipper of anthracite, attempted to create several anthracite marketing combina-

This portion of the map of the Lehigh Valley Railroad ("The Route of the Black Diamond") shows the large number of railroads in the area of the anthracite mines, leading to the main line. Many iron furnaces and industries that relied on anthracite for steam power were founded along the route.

The railroad extended to tidewater and to the Great Lakes. (NCM/D&L)

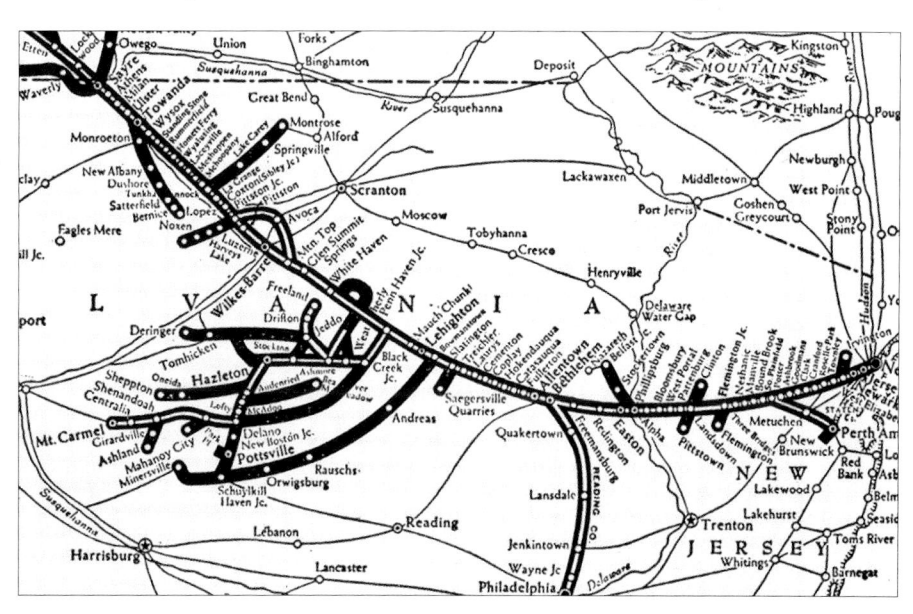

tions during the 1870s and 1880s. These attempted to apportion market share and limit production. The energetic and ruthless Gowen made the P&R the largest corporation in the country by the early 1870s, but his overreach led to the line's first bankruptcy in 1880.

In 1890 Archibald A. McLeod became president of the Philadelphia and Reading Railroad. McLeod was a hard-boiled and extremely ambitious businessman who planned not only to dominate the anthracite-mining and -transportation industries, but also to transform his railroad into one of the dominant lines in America. During 1891–92, McLeod gained control of the cash-poor Lehigh Valley Railroad and Central Railroad of New Jersey by leasing their lines and guaranteeing their profits out of the Philadelphia and Reading's earnings. McLeod further consolidated the Philadelphia and Reading's control over the anthracite industry by buying control of the Delaware, Lackawanna and Western Railroad. By the spring of 1892, his empire controlled 72 percent of anthracite-production facilities in America. By August of 1892 he had gained control of the Central New England and Western Railroad, and the Boston and Maine Railroad. These acquisitions made the Philadelphia and Reading a major trunk line, but by February of 1893 McLeod's empire had collapsed because the earnings of the over-burdened P&R could not support the inflated payments that were due to the owners of the various leased railroads.

The Philadelphia and Reading once again fell into bankruptcy and McLeod was ousted. Many economic historians contend that its failure triggered the devastating Panic of 1893. Seizing the opportunity, powerful New York investment banker J.P. Morgan took over the railroad; by 1900 he had also gained control of the other major anthracite lines. Morgan believed that unfettered competition was both wasteful and unnecessary, and he created what he termed a "community of interest" among the major anthracite railroads as a part of his master plan for the consolidation of eastern railroads. Under Morgan's auspices, the Philadelphia and Reading gained permanent control of the Central Railroad of New Jersey and was itself controlled by the Baltimore and Ohio. However, opposition from the federal government prevented full implementation of Morgan's desires to create further mergers.[8]

Cement Spreads

The 1890s saw a dramatic expansion of the manufacture of portland cement in the Lehigh region. Quarries began to be opened up along the length of the Jacksonburg formation. American Cement in Egypt grew from one mill to four, and opened up vast quarries that still can be seen in Whitehall Township. Cement kilns were constructed by new companies such as Dragon and Bonneville in Siegfrieds, Whittaker and Vulcanite in Alpha, New Jersey, and Nazareth Cement in Nazareth. New towns—Ormrod and Cementon—sprang into being around the new Lehigh and Whitehall cement companies.

In 1896, José de Navarro's Atlas Cement opened a plant in Stemton (one of three villages that later became Northampton), across the Lehigh River from the original Keystone "Valley Plant" rotary kiln (which kept operating) in Coplay. The last of the vertical stationary kilns, the Schoefer kilns, were built in 1893 by Coplay Cement, but they were almost instantly obsolete as the hori-

The Lawrence Portland Cement Company, later called the Dragon Company, was the only cement company to transport cement on the canal system. All other cement companies moved their product by rail.

Here, one loaded boat is leaving the loading chute at Siegfried while a second is moving its front section forward to be filled. Like coal boats, these were section boats with two parts coupled together. Each section was loaded and unloaded after uncoupling; this maintained some equilibrium.

(NCM/D&L)

Only one of the original ten Schoefer kilns appears to be in operation in this circa 1900 photograph of Coplay Cement Mill B. Though these were state-of-the-art vertical cement kilns when they were built in 1893, they were already being out-produced by Navarro's nearby horizontal rotary kiln.

Eight of the kilns, now the only ones extant in the world, are still standing in Saylor Park in Coplay. The site is on the National Register of Historic Places.

(Private collection)

zontal rotary kiln became dominant. The lower parts of the Schoefer kilns still stand on the former site of Coplay Cement, now a park near the Lehigh River in the borough. Their tops were removed in the 1940s for safety; their present appearance is very unlike how they looked when they were in operation.

The Atlas—later the Universal Atlas Cement Company—grew to become the biggest cement plant in the world, and the source of virtually all the cement used in the construction of the Panama Canal. The presence of the burgeoning cement plant led the three villages along the stretch of the Lehigh Canal—Stemton, Newport, and Siegfrieds—to form an "alliance" in 1902; in 1909, the cement company petitioned the county courts to change the alliance into a borough called Northampton, largely to make it easier for the firm to send and receive mail.

Lehigh Cement was founded in 1897 by six Allentown businessmen led by Harry Trexler, who had made fortunes in lumber and farming. Trexler, Edward Young, and George Ormrod invested most of the $250,000 it took to construct a cement plant at the village of Ormrod. Lehigh Cement soon established a second facility at West Coplay, followed by another plant in Ormrod. Since the company's cement was being shipped as far west as Kansas City, another plant was built in Mitchell, Indiana, in 1902. The following year a third cement plant was constructed in

Ormrod, and in 1906 a second, larger plant was built in Mitchell. Lehigh Cement built a plant in Fogelsville, Lehigh County, in 1907, and a Mason City, Iowa, facility—the company's first venture west of the Mississippi—in 1911.

Though there were cement plants in operation in other regions of the country, in 1890 the Lehigh District, as the area was known in the industry, produced 60 percent of the cement made in the U.S. By 1900, Lehigh District kilns turned out over 16 million barrels of cement, nearly 75 percent of the annual production of cement in the entire United States. This output made the Lehigh Valley the greatest cement-producing region in the world.[9]

Angels in the Anthracite Regions

A paradox of the Industrial Revolution was that it drove millions of men, women, and children into tedious, exhausting, and poorly paid work in hazardous working conditions, while simultaneously producing more wealth spread among more people than ever before. That wealth gave rise to philanthropy, which not only endeavored to relieve working people's immediate suffering and poverty, but to address the causes. Though industrialists like the men who built the industries in the Corridor made the money, it was often their wives who had the time, education, and religious zeal to do the work of doing good.

A central tenet of nineteenth-century philanthropy was that education was the path out of poverty, by imparting not only knowledge and useful skills, but self-respect, self-discipline, and morality. Two women who embodied that belief were Wilkes-Barre's Ellen Webster Palmer, and Drifton's Sophia Georgina Fisher Coxe.

Ellen Palmer (1839–1918) was born in Plattsburg, New York, and trained as a teacher. Her husband, Henry W. Palmer, was a lawyer who served as Pennsylvania's attorney general from 1879 to 1883. When the couple returned to Wilkes-Barre, Ellen Palmer became concerned about the terrible working conditions of breaker boys, and their complete lack of both education and healthy recreation. She started hosting Saturday evening entertainments for the boys, many of them only nine or ten despite the passage in 1885 of laws prohibiting boys under 12 from working in the breakers. Her efforts expanded to storefront classrooms where the boys learned arithmetic, reading, and writing. Eventually, she organized the Boys Industrial Association (BIA) and tapped wealthy friends for financial support.

Ellen Palmer with a group of the breaker boys whose lives she worked to improve through education.

(From the Photograph Collection of the Luzerne County Historical Society)

The goal of the BIA was to set boys on a path to work other than in the mines. Her son, Bradley Webster Palmer (later a founding partner in Gillette Company and International Telephone and Telegraph), endowed a scholarship that enabled boys to continue their education, even after the BIA ended with Ellen Palmer's death. A statue of her, "The Friend of the Working Boy," stands near the Luzerne County Courthouse.[10]

Sophia Georgina Fisher Coxe (1841–1926) was born in Philadelphia to a wealthy family, and was educated there and in Switzerland. In 1869, she married Eckley Brinton Coxe, grandson of Tench Coxe, one of the pioneer investors in anthracite. Eckley himself was, at age 29, already president of the largest independent anthracite company in Pennsylvania. Not intending to manage from afar, Eckley built a house for his bride in Drifton, Luzerne County. There Sophia came face to face with people, cultures, and a level of poverty that were a far cry from her privileged high society life in Philadelphia. But Sophia's strong will and overwhelming sense of duty, which had been notable since childhood, paired with the couple's devout Episcopalian faith, guided her life in the anthracite region.

Eckley and Sophia soon built a community center in Drifton for meetings, entertainment, education, and cultural programs, a library, and a technical school to teach safe and efficient mining processes. Sophia organized large annual community July 4th celebrations that featured entertainment, fireworks, and lavish food. Every year in early December, railroad box cars arrived in Drifton loaded with practical goods like blankets, warm coats, hats, and boots as well as toys and trinkets. Sophia personally selected the gifts for all 3,000 of the mine company's workers and their families, and kept meticulous lists so that no one got the same gift the following Christmas.

The Coxes established the first hospital in the area, an offshoot of Sophia's earlier practice of driving a wagon around the patch towns with medicine for whoever needed it. From a trip to Europe, they brought back a cure for usually fatal diphtheria, which was said to have eliminated the disease in the mine towns. Sophia and Eckley's most lasting achievement was founding "The Mining and Mechanical Institute of the Anthracite Coal Region of Pennsylvania," better known as MMI. At its beginning in 1893, it provided technical education to miners and mechanics who maintained mining equipment, with the aim of teaching young men with demonstrated talents and work ethic the advanced skills that would enable them to manage mines, miners, and their equipment efficiently and safely. The school soon added college prep courses that set the most promising young men on the path to higher education. Eckley died shortly after the school was founded, but Sophia maintained

The only known photograph of both Sophia and Eckley Coxe. They are on the porch of their home in Drifton.

(Courtesy of MMI Preparatory School, Freeland, PA)

their financial support of the school, which, by the time of her death in 1926, had graduated 217 students, of whom 135 went on to college, 59 of those to Lehigh.

The MMI Preparatory School, now an independent, private, co-ed school for students from grades 6 to 12 from the Freeland area, offers a rigorous, multi-disciplinary curriculum. The Sophia G. Coxe Charitable Trust maintains its offices in their former home in Drifton, funding more than thirty social services agencies in the immediate anthracite region.[11]

Zinc Moves to Lehigh Gap

Despite the steady decline of the Saucon Valley mines during the 1880s, the Lehigh Zinc Company continued to prosper by importing ores from other locations. It became a major producer of spiegeleisen, a low-grade form of ferro manganese that was an essential ingredient in the production of Bessemer steel. By 1877 its works occupied ten acres of land east of New Street and north of Second Street in South Bethlehem. It employed over four hundred men and contributed much to the community's benefit, such as support, along with that of Asa Packer and the directors of Bethlehem Iron, for the creation of St. Luke's Hospital.

The Lehigh Zinc Company's plant at South Bethlehem, circa 1875. (Library of Congress)

In 1881 the Lehigh Zinc Company was purchased by Samuel Wetherill and his partners, John P. Wetherill, Richard B. Wetherill, and August Hecksher. Reorganized as the Lehigh Zinc and Iron Company, the company's management was dominated by John Wetherill, who became its general manager. Under his capable direction the works increased in size and by 1892 it was producing 1 million pounds of metallic zinc and 1.2 million pounds of zinc oxide per month.[12] In 1896 John Wetherill invented a magnetic separator for concentrating zinc ore. He also purchased, along with his brother Samuel, a tract of land near Lehigh Gap as a possible site for an enlarged works.

In 1897 a major consolidation of the zinc industry occurred when the New Jersey Zinc Company was reorganized. Descended from the Sussex Zinc and Copper Mining and Manufacturing Company, this concern had established a national reputation for the quality of its metallic zinc products, which were marketed under the "Horsehead" brand name. In 1897 it absorbed almost all of its competitors, including the Lehigh Zinc Company, and in 1899 it started to acquire all the mine sites in the Saucon Valley. Having created a virtual monopoly of America's zinc industry, the New Jersey Zinc Company decided to consolidate all of its scattered zinc smelting plants at one location and selected the land purchased by the Wetherills near Lehigh Gap for the site.

The Lehigh Gap area had long been a focus of commercial activity along the Lehigh Navigation and the Lehigh and Susquehanna Railroad. Located almost midway between Bethlehem

Craig's Hotel in the Lehigh Water Gap, and the covered bridge over Big Creek. The earliest European settlement at the gap was here. Sketch by Rufus Grider, July 1853.

(Courtesy of Moravian Archives)

and the anthracite coalfields, the Lehigh Gap community developed around a general store and tavern owned and operated by the Craig family. A small station called Hazard, on the Lehigh and Susquehanna Division of the Central Railroad of New Jersey, was at Lehigh Gap. The area had much to offer as the site for a zinc smelter, since it was close to the anthracite coalfields, which would provide it with fuel, and had good railroad connections to the zinc mines at Franklin and Ogdensburg in New Jersey.

Development of the new zinc manufacturing facility and its associated community began in 1896-1897 with the formation of the Palmer Land Company. This wholly owned subsidiary of the New Jersey Zinc Company was named after Samuel S. Palmer, the company's president. In 1898 ground was broken for a plant at Lehigh Gap along the Lehigh River, and by the end of that year a powerhouse, oxide furnace, spelter-making facility, and office building had been finished. These structures became the basis for what later was termed the West Plant at Palmerton. As the new century dawned, the company also began to plan and build a model industrial community—Palmerton.[13]

By the time the New Jersey Zinc Company built its first plant at Palmerton (in the distance in this photo), the area was already well served by railroads. The tracks of the Central RR of New Jersey and the Lehigh Valley RR were on opposite banks of the Lehigh River. The Lehigh and New England came through at a higher level, with an impressive bridge at Lehigh Gap. The Chestnut Ridge Railway opened in 1900, failed after a few years, and was taken over by the zinc company. It had interchanges at Palmerton with the larger CNJ and L&NE railroads.

(NCM/D&L)

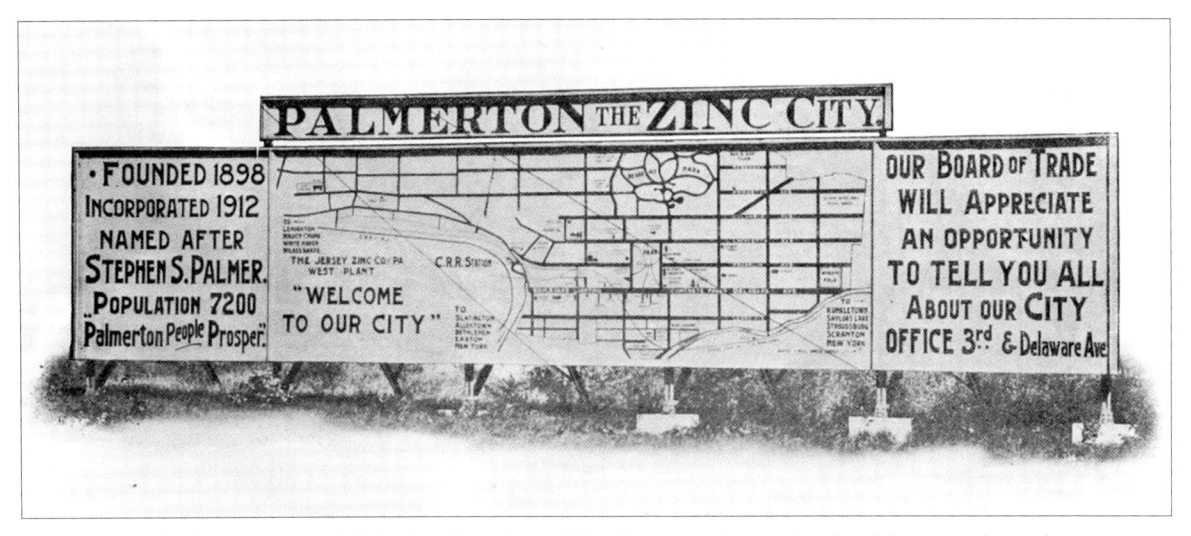

This image was published in New Jersey Zinc Company's 1923 book celebrating the 25th anniversary of the founding of the town and the plant.

New Jersey Zinc Company showed a concern for its workers that was unusual for its time and for businesses in the United States. While building Palmerton and the zinc works from the ground up, the company's social policies were concretely manifested. In 1907, it established a "Sociological Department" and opened a Neighborhood House. There residents—whether or not they worked for the company—could take advantage of a wide array of social, cultural, educational and recreational opportunities for all age groups. The second Neighborhood House, shown here circa 1918, is now Palmerton's borough hall.

(Courtesy of Aaron Heckler)

Chapter Eleven

1900 to 1910
New Century, New Industries, and a New Name

*"I intend to make Bethlehem the prize steel works of its class,
not only in the United States, but in the world."*

Charles M. Schwab, 1904

Palmerton, the town that zinc built, was planned as a model industrial community with unusually wide streets, spacious workers' housing, a citizens' cooperative, a central park, and modern utilities. Architect William E. Stone of New York City designed the first fifty workers' homes, four superintendents' residences in a park on the north side of the town, the company store, and the company hotel that became known as the Horse Head Inn. Homebuilding took place in phases as the company expanded, several hundred new houses being built some years.

The Palmer Land Company sold its houses to the workers of the New Jersey Zinc Company on an installment plan. Each prospective homeowner had to make a down payment equal to 10 percent of the dwelling's value, after which equal payments would be deducted from his pay during each of the next 43 months. At the end of this period he would have paid 35 percent of the purchase price and would be granted a five-year mortgage at 4 percent annual interest in order to pay off the remainder. Over 393 workers had purchased their own homes under this system by 1923.

During the first decade of the twentieth century, the quality of life at Palmerton greatly improved as the Palmer Land Company built schools, a well-designed park, a community center, and a hospital. A citizens' cooperative provided activities for all ages, and the Neighborhood House, sponsored by the company, provided amenities such as kindergarten for the children of workers, and baths for the poorest immigrant workers. The present Borough Hall is the second Neighborhood House, built to be an all-purpose municipal center. The company arranged for the Central Railroad of New Jersey to stop for passengers in Palmerton instead of at Hazard in Lehigh Gap. The stone railroad station, built in 1911, is one of finest remaining in the Corridor.

The Palmer Land Company encouraged other industries to locate in the community; in 1903 the Read and Lovett Silk Manufacturing Mill was established to provide employment for the wives and daughters of zinc workers. By 1907 the West Plant had grown to such a size that the old

Lehigh Zinc works at South Bethlehem could be demolished. The operations of the New Jersey Zinc Company expanded at Palmerton with the construction of a new plant on Palmerton's east side. Begun in 1908, the East Plant by 1913 rivaled the long-established West Plant in size and productivity. Palmerton's early development culminated in 1912 when it became a separate borough.

The West Plant by 1912 comprised facilities for the manufacture of zinc oxide, zinc slabs, and spiegeleisen. The East Plant, originally built to provide increased production capacity for zinc oxide used in the manufacture of rubber tires, was greatly enlarged during World War I. After the war it also produced zinc-based pigments, luminescent pigments, sulfuric acid, metal powders, ferro manganese, and rolled zinc. During the 1920s the world's largest zinc-related research laboratory was built at Palmerton after the New Jersey Zinc Company decided in 1919 to centralize all of its research activities in a new central laboratory complex, choosing Palmerton as its site.[1]

The far west end of the West Plant of New Jersey Zinc in Palmerton, the first of two zinc processing plants along the river. The photo was taken probably around 1930 by noted Carbon County aviator Jake Arner, whose name is memorialized in a small airport at Lehighton. This is probably a Sunday because so many canal boats are moored, and there is no smoke coming out of the smoke stacks.

The canal boats in the photo had brought coal to fuel the spiegeleisen furnaces. Spiegeleisen was a low-grade form of ferro manganese that was an essential ingredient in the production of Bessemer steel. It was made from the clinker left over from zinc oxide production, and the zinc plant produced it exclusively for Bethlehem Steel.

The canal was later filled in and Pennsylvania Route 248 was built over it.

(NCM/D&L)

The Palmerton Hospital, built by New Jersey Zinc, was the first one in Carbon County. Opened at its present Lafayette Avenue site in 1908, the hospital was enlarged in 1911 and regularly during the next two decades.

During the 1920s, Palmerton Hospital was one of the first institutions of its size to be accredited by the American College of Surgeons. The hospital became part of St. Luke's University Hospital Network in 2018.

(Courtesy of Aaron Heckler)

Delaware Avenue, Palmerton's main retail and business street, is an unusual ninety feet wide. It was built that way in anticipation of construction of an Allentown-to-Stroudsburg trolley line that never materialized.

One mile was paved with concrete very early, some time before the 1923 anniversary.

(Courtesy of Aaron Heckler)

Delaware Avenue looking east, showing the wide street lined with early twentieth-century shops.

(Courtesy of Ron Bray)

The progressive managers of New Jersey Zinc included a green nine-acre park in the heart of their town.

Now officially known as Borough Park, it has long been a center for community activities and restful recreation. For at least a century it has boasted a classic, small-town bandstand.

This postcard was mailed in 1911.

(Courtesy of Ron Bray)

Mack Trucks moves to Allentown

Mack Brothers Motor Car Company was incorporated in Lehigh County on January 2, 1905. Brothers Jack, Gus, and Willie Mack had built a successful wagon- and carriage-making business along Atlantic Avenue in Brooklyn, New York, during the financially fraught 1890s. They had then moved into building gasoline-engine open buses for taking sightseeing tourists around Prospect Park and Manhattan.

The move to Allentown from New York was driven by a need for more space to build buses than their original factory allowed. It also brought them back to their home state—the brothers had all been born on the family farm near Mount Cobb in Lackawanna County. One brother, Joseph, was already in Allentown, running two silk mills with partner Leo Schimpff. Near one of them, on the banks of the Little Lehigh Creek, was the large empty foundry and machine shop of the Weaver-Hirsh company, which had folded in 1904. The Macks moved their bus business there, and then began to work on developing motor trucks. In many ways, this was the real driving force behind the move to Allentown, for they knew that they would not have had enough space to build all the components—chassis, body, engine, driveshaft and transmission—in their Brooklyn workshops.

Selling a new and unproven product like a motor truck meant doing a lot of marketing to an often skeptical business audience, and the Macks found that many customers insisted on a trial period of their new vehicle before they would commit to buying outright. This meant that the company was spending money on labor and expanding its facility at a time when there was not a lot of cash flowing back in. To stay afloat, Jack Mack tirelessly pursued investments by Allentown's prominent businessmen, while brother Joe, still in the silk business, persuaded his fellow silk men—who were becoming more numerous every year—to invest in trucks as well.

Mack's new trucks began to replace the horse-drawn delivery wagon, but slowly, before 1910. This 5-ton truck is loaded with 100-lb sacks of flour, more than one draft horse could easily haul.

For a few decades after the introduction of motorized trucks, horse-drawn wagons continued to be used, most often for door-to-door daily deliveries of milk. The horse would memorize the route, leaving the "driver" free to run up to each front door with bottles of milk and cream.

As city streets began to be paved, the purchase of a motor truck became much more attractive to local businesses. In 1909 Mack introduced a lighter-weight truck that became very popular. *(Courtesy of Mack Trucks Historical Museum)*

A group of 63 workers at the spacious new Mack plant in Allentown in 1907.

As the infant truck industry grew and evolved, several mergers took place. Mack Brothers and the Saurer Motor Company merged in 1911 to become the International Motor Company, with Hewitt trucks also part of the combine.

The Mack name is the only one of the very early truck manufacturers to survive.

(Courtesy of Mack Trucks Historical Museum)

One of Allentown's attractions was the large labor pool of skilled machinists, mechanics, foundry workers, blacksmiths, wheelwrights, draftsmen and other workers. The Mack brothers realized that men with such in-demand skills could and would pick up and move to another job at the first sign that the new company was on shaky financial ground. So they paid fair wages, even when the company cash flow was low, and organized small celebrations on pay days and bigger ones when a truck sold. Company lore recalls that Jack Mack would lead a group of workers to a pretzel "foundry" on 10th Street, and then on to the adjacent brewery to quench their salt-induced thirst.[2]

The first year in the new factory "proved to be a highly formative one for the Mack product, with most of the major component innovations retained as standard designs through 1912."[3] This included a four-cylinder gasoline-powered engine that made 50 to 60 horsepower. These were installed in a 3/4-ton truck known as the Manhattan.

By 1909, motor trucks had crossed the line from novelty to necessity for many businesses. But most did not have the need for such heavy-duty models as the Manhattan, so in 1909 the Macks introduced a 1-ton truck made with a lighter aluminum frame and powered with a smaller 32 hp motor. This model quickly became known as the Junior. Dump trucks and tank trucks, to deliver the petroleum and gasoline that was rapidly replacing coal as the nation's fuel, soon followed. Demand for trucks soared all over the country, which not only drove the Mack Brothers' business to an even faster growth rate, but created a vast number of competitors.

The engines for the Junior trucks were purchased from a small Newark, New Jersey, firm called F.A. Seitz. Faced with the possibility of not being able to source enough engines because of the rising competition, the Mack Brothers Motor Car Company purchased the Seitz operation in order to assure a steady supply of engines. About the middle of 1910, the "Manhattan" name was dropped and all the motor vehicles made in Allentown had the name "Mack" on the sides of their cabs.[4] The iconic bulldog mascot was still more than 20 years in the future.

Bethlehem Iron Becomes "The Steel"

The organization of the Bethlehem Steel Company on April 17, 1899, marked the end of an era in the history of iron and steel making at Bethlehem. Of the many anthracite-fueled furnaces in the Corridor, Bethlehem Iron was the only one to move to making steel. Beginning with the retirement of John Fritz in 1893, continuing with Robert H. Sayre's retirement in 1898, then the departure in 1902 of Russell W. Davenport to assume management control of the Cramp Shipbuilding Company in Philadelphia, all the principal figures in the company's early development ended their involvement with the company. Even before Davenport's departure, ownership of the Bethlehem concern had changed hands.

In 1901 the British firm of Albert Vickers Sons and Maxim, wanting to enter the American ordnance and armor market, made an offer to buy control of Bethlehem Iron. Vickers was an immensely powerful conglomerate that produced armor plate, cannon, and warships, and held the principal patents for the self-acting machine gun. On May 28, 1901, Vickers offered to purchase all of Bethlehem's stock at a price of $22.50 per share. The Vickers offer was rejected and a counter offer of $24.00 per share from Charles M. Schwab, president of the newly organized United States Steel Corporation, was accepted on May 30, 1901. On August 15, the stockholders of the Bethlehem Iron Company voted to accept Schwab's offer. Bethlehem Iron Company's stockholders received a $1,000 bond for every twenty shares of company stock. The lease between the iron company and the steel company, which had been created two years before when Bethlehem Steel was organized, was canceled and the iron company ceased to exist. In its place was a transformed Bethlehem Steel Company with a capitalization of $15,000,000 and complete operational control of the Bethlehem plant.

Joseph Wharton

In 1904, Charles M. Schwab and Joseph Wharton formed the Bethlehem Steel Company out of the former Bethlehem Iron Company. This took control of the business almost completely out of the hands of the local, original owners and directors. Schwab led the company into the production of structural steel and, at great financial risk, installed a revolutionary technology for producing the continuously rolled wide-flange beam. By 1910 the plant in South Bethlehem was the largest individual steel plant in the country. It spread over 1,500 acres and employed more than 9,000 workers (compared to 1,000 in 1870), at a time when the entire population of South Bethlehem was about 25,000.[5]

Charles Schwab was in his early 40s when he took charge of Bethlehem Steel as president and chairman of the board.   (NCM/D&L)

A Melting Pot Called South Bethlehem

Perhaps as many as half of the workers were immigrants. The foreign-born percentage of the borough's population rose from about 30 percent in 1870 to about 45 percent in 1900. The 1910 U.S. Census identified people from 52 different nationalities in South Bethlehem. Joan Campion wrote in *Saturday Night on the South Side*:

> In its heyday, this area must have looked—and especially sounded—like the world in a small space ... Outside of large metropolitan areas, there must be relatively few neighborhoods as small as the South Side where so much ethnic diversity could be found, and where so many languages could be heard.[6]

In the usual pattern of immigration to the U.S., a very high proportion of these later immigrants from Hungary, Poland, Czechoslovakia, Italy, Greece, Russia, and the Ukraine were single young men rather than entire families. They clustered in the neighborhoods largely made up of boarding houses and saloons that were rapidly spreading eastward along with the expanding steel mill. By 1900, South Bethlehem encompassed twice its original area—and had a growing reputation as a "sin city."

It was no wonder, with all those single men with some money in their pockets, that bordellos and gambling halls flourished. Many out-of-towners from as far away as New York and New Jersey came to South Bethlehem to have some illicit fun in a place where no one would recognize them. When Prohibition began, speakeasies added to the attractions of the South Side.

In 1924, the noted journalist and curmudgeon H.L. Mencken and his friend and publisher Alfred A. Knopf came to Bethlehem for the Bach Festival, described by Campion as "incongruously perched on the far side of what was in fact a red light district." Hoping for a bit of liquid refreshment after the concert, Mencken and Knopf were taken to a speakeasy by their cab driver. They were denied entrance until Mencken, waving the score of the B Minor Mass they had just heard, convinced the doorkeeper that they were musicians, not federal agents.

Despite much empirical evidence that many of the perpetrators of unsavory behavior and even outright crime on the South Side were from out of town rather than residents, city politicians

By 1910, South Bethlehem was home to people from 52 nationalities. Most of the workers rented rooms, or even just a bed in a rooming house. Unlike many industries in the Corridor, Bethlehem Iron, and later Bethlehem Steel, built few company houses for its employees. This row, in the 1200 block of East Third Street, is still standing and occupied.

(NCM/D&L)

and many residents of the other parts of the city laid all the blame for the goings-on across the river on the "foreigners." Beginning in 1929, some reform-minded mayors pledged to "clean it up"—but that was a long, slow process, says Campion, culminating in the urban-renewal-driven destruction of much of the South Side's architectural and ethnic character.[7]

Steel, Ships, and Schwab

Charles M. Schwab (1862–1939) was among the most brilliant and innovative steel makers in America. As the protégé of Andrew Carnegie, he had risen to become president of Carnegie Steel; he personally negotiated the merger of Carnegie Steel with the steel interests of J.P. Morgan to create the United States Steel Corporation, of which he became the first president. Schwab purchased Bethlehem as an independent investment, but soon thought better of it and transferred control to United States Steel. However, when approached by a group of investors seeking to create a shipbuilding conglomerate, Schwab was able to repurchase the Bethlehem Steel Company for $7,246,000; he then transferred it to the newly organized United States Shipbuilding Company in return for a large interest in the new concern. He also placed limits on the amount of control that the United States Shipbuilding Company could exercise over Bethlehem.

The United States Shipbuilding Company almost immediately began to experience financial difficulties. Although some of its shipyards were modern and efficient, many others were old, obsolete, and had almost no customers for their limited products. In contrast, Bethlehem Steel Company was a prosperous enterprise. Its forging plant with its large-capacity machine shop was regarded as the finest in America and was recognized as such in 1902 when it was chosen to build a 12,000-ton forging press and pumping engine for its principal competition, the Homestead Plant of the United States Steel Corporation. To produce these great forging devices, Bethlehem produced the largest and heaviest steel castings yet made, some of which weighed more than 325,000 lbs.

During this period, the Bethlehem plant employed more than 9,461 workers and dominated the American ordnance and armor market, despite competition from Midvale Steel, which also went into armor production. Partly because Schwab refused to allow a major proportion of Bethlehem's profits to be diverted to the parent corporation, the United States Shipbuilding Company failed in 1903. Despite a series of acrimonious lawsuits, Schwab was able to regain complete ownership of the Bethlehem Steel Company while at the same time salvaging the stronger of the United States Shipbuilding Company's shipyards. He combined these properties into the Bethlehem Steel Corporation, which was organized on December 10, 1904.

The new concern, of which Schwab immediately became president, was capitalized at $15,000,000. Its primary properties were the Bethlehem plant, the Harlin and Hollingsworth shipyard at Wilmington, Delaware, the Union Iron Works shipyard at San Francisco, California, Samuel L. Moore Son's Co. ship-repair yard at Elizabethport, New Jersey, the Easton Shipbuilding Co. of Groton, Connecticut, the Crescent Shipyard at Elizabethport, New Jersey, the Bath Iron Works Shipyard and the Hyde Windlass Company, both at Bath, Maine, and the Carteret Improvement Company, which controlled a large parcel of undeveloped land at Carteret, New Jersey. Many of these shipyards proved to be of little value. By 1907, the Bath Iron Works, the

Hyde Windlass Company, and Eastern Shipbuilding Company had been sold and the Moore and Crescent shipyards were consolidated. Despite these divestitures, Schwab had created the basis of a greatly enlarged company.

Even before the Bethlehem Steel Corporation was formally organized, Schwab stated his great plans for it:

> I intend to make Bethlehem the prize steel works of its class, not only in the United States, but in the entire world. In some respects, the Bethlehem Steel Company already holds first place. Its armor plate and ordnance shops are unsurpassed, its forging plant is nowhere excelled and its machine shop is equal to anything of its kind. Additions will be made to the plant rather than changes in the present process of methods of manufacture.[8]

Schwab planned to use the Bethlehem plant as the centerpiece for a large steelmaking concern that could compete successfully with United States Steel Corporation, the presidency of which he had been forced to relinquish on August 4, 1903. Schwab believed that a steel company should be aggressively managed to continually seek ways to cut costs and prices, with a resulting increase in market share. This business philosophy clashed with the more conservative attitudes of Judge Elbert Gary, chairman of the board of U.S. Steel; in the ensuing power struggle Schwab had been forced out. Since Schwab had helped create U.S. Steel, he knew intimately that concern's strengths and weaknesses, and he planned to exploit the market opportunities that were created by its conservative management policies whenever possible.

Schwab realized that Bethlehem's continued dependence on military contracts was in the long term dangerous to the company's continued prosperity. Bethlehem continued to dominate the United States market for such items as ordnance forging, but Schwab realized that the new corporation had to develop an expanded line of civilian products to cushion the shock of a sudden downturn in U.S. government orders. Among his first acts, he ordered the installation of a crucible steel plant to produce high-quality alloy steels, and an open-hearth-steel rail mill. Since the U.S. Steel Corporation did not have a similar facility, Schwab could charge a premium for the superior products made in the new facility and, since his competitor was loath to scrap its large investment in Bessemer-steel rail mills, Bethlehem faced little threat in the rail market from its much larger rival.

Schwab also extended the range of Bethlehem Steel Corporation's forging activities by adding a large-capacity drop-forging operation during 1905. Bethlehem's drop-forge facility soon won a reputation for the high quality of its products and during the 1920s it pioneered in the production of forged cylinders for the newly developed radial air-cooled aircraft engines. Until the 1980s the drop-forge operations were a major profit center for Bethlehem Steel.

The heavy forging facility is the only part of the enormous Bethlehem Steel plant that is still operating. Lehigh Heavy Forge Corporation, a subsidiary of WHEMCO, Inc., is the only remaining super-heavy forging plant in North America.

The earliest known view of the Grey Universal Beam Mill, circa 1910. The wide-flange or H-section structural beams and columns produced by this mill revolutionized construction of taller skyscrapers and longer bridges. They were quickly adopted by engineers and architects when building rebounded after the Panic of 1907.

Bethlehem Steel was the sole producer of H-beams in the United States until 1927. That year, after a patent-infringement suit with U.S. Steel was settled, Bethlehem granted its rival a license to produce the beams.

(NCM/D&L)

Building Beams

Schwab's boldest decision was to enter Bethlehem into the growing but fiercely competitive structural steel market. Once again, his knowledge of the strengths and weaknesses of the United States Steel Corporation was a decisive factor in his planning. He knew that the Homestead Plant of U.S. Steel was among the largest and most productive manufacturers of structural steel shapes in America and that it would be foolhardy for Bethlehem to compete directly with this colossus. As he had done with the installation of an open-hearth rail mill, Schwab planned to produce a superior product to fill a new market niche, one that was uncontested by U.S. Steel. He found his product in 1905 when he committed Bethlehem Steel Corporation to the production of the continuously rolled wide-flange ("Grey") beam.

Immigrant British engineer Henry Grey (1846–1913) had developed a revolutionary process to roll beams while he was at the Ironton Structural Steel Company of Duluth, Minnesota. In 1902 he installed his first full-scale structural mill at Differdingen Steel Works in Luxembourg. Grey's mill could roll wide-flange beams continuously, directly from ingots. The beams had wider flanges, making them "H" shaped instead of "I" shaped, thus were stronger and less likely to bend than conventional beams of the day. They were also much cheaper both to produce and to use because they were rolled continuously as a single section, and the high costs of using conventional beams (riveting and other fabrication) were eliminated.

Schwab learned of Grey's successful innovation while he was president of U.S. Steel, but corporate chairman Judge Elbert Gary did not believe that Grey's beam could be mass produced, and had rejected it. Schwab secured the rights to Grey's invention for Bethlehem in 1908.

The decision was a bold move and a potentially dangerous financial gamble for the Bethlehem Steel Corporation. An investment of almost $5,000,000 would be needed to build the new division of the Bethlehem plant to produce the wide-flange Grey beam. During the next year Schwab attempted to raise this sum from a variety of sources and by July of 1908 the Saucon Division of the Bethlehem plant, with its open-hearth furnaces and structural mill, was placed in full operation.

Development of the Saucon Division was largely the responsibility of Eugene Grace (1876–1960), who became Schwab's chief protégé and his eventual successor as the head of Bethlehem Steel. However, although Grey beams were being successfully produced they found few buyers. Schwab was forced to turn to Bethlehem's forging operations for financial salvation.

Since domestic orders for ordnance and armor plate were flat, Schwab hoped to increase foreign sales of military hardware. For more than a decade Bethlehem had been a major factor in the international arms and armor-plate market and was a charter member, along with Krupp, Schneiders, and Vickers-Armstrong, of the international armor-plate cartel, the Harvey United Steel Company, Ltd. This was a patent pool that held almost all of the important patents for armor-plate production; it also served as an informal way for its members to divide up the market. Schwab sought to increase Bethlehem's share of this lucrative trade to get the funding needed to subsidize operation of the Grey mill. Archibald Johnston, Bethlehem's vice president of sales, had long been associated with the forging operation, and was sent to Europe to negotiate with the other members of the Harvey United Steel Company. He was successful and by 1908, Bethlehem's share of the international armor-plate market had risen to a level of $2,000,000 annually.

With profits coming from the increased foreign military sales, Bethlehem was able to continue marketing the Grey beam. By 1909 the beam had entered the marketplace, ensuring the success of Schwab's efforts to diversify Bethlehem's product line. Architects and engineers saw the value of the product, and by 1914 sales of structural steel were double the annual total value of Bethlehem's forging sales. At the same time, the U.S. government began to increase its orders for armor plate and ordnance forgings. In 1909, Bethlehem received an order for 7,731 tons of armor from the U.S. Navy, valued at $2,300,000, and in 1910 it received its largest single order to date for military products when Argentina purchased ordnance, shells, and armor valued at more than $10,000,000. Orders such as these continued to keep the forging operations at Bethlehem in full operation.[9]

Bristol Prospers

By 1900, Bristol was the manufacturing center of Bucks County. Through the 1880s and 1890s, the Bristol Improvement Company had continued to build industrial buildings and attract both new and expanding businesses to join the ever-growing Grundy Woolen Mills.

In 1886, the Thomas L. Leedom Carpet Company of Philadelphia moved into the last BIC-constructed building, which became known as the Bristol Carpet Mill. Leedom produced Wilton rugs, high-quality, tightly woven carpets that were considered to be the finest of their kind made in the United States. Thomas L. Leedom vertically integrated his operations after relocating to Bristol, incorporating weaving, dyeing, yarn spinning, and, eventually, wool preparation under a

The Leedom Carpet Company's Bristol Carpet Mill, a major employer in the town.

(Courtesy of the Grundy Library)

single roof. By the early twentieth century, Leedom was the second-largest employer in Bristol, with over 850 workers.

The smaller mills erected by the BIC never attained a level of industrial or economic importance to equal the Grundy and Leedom operations. The Keystone Mill, which started as a fringe manufacturing plant operated by L. M. Harned & Co., became a warehouse for the Grundy Mill in 1885. In 1903 Edward T. Steel & Co., a manufacturer of men's worsted cloth that had taken over the Livingstone Mill in 1887, purchased the property and incorporated the building into its worsted mill. The Star Mill passed through the hands of three owners before being annexed to the adjacent wallpaper mill in 1891. The wallpaper mill operated intermittently, under the control of at least four different firms, throughout the late nineteenth and early twentieth centuries. Other industries, including leather tanning and patent-leather production, and a foundry, came to Bristol in the 1890s and the early years of the twentieth century. The Grundy mill was by far the biggest and most successful concern.

The Grundy Mills consisted of four large, connected buildings, with a total of over 300,000 square feet. The railroad track in this postcard view allowed railcar delivery into a courtyard in the middle of the complex.

The mill was the largest employer in Bucks County; thirty percent of all the workers in Bristol were employees of the firm.

(Courtesy of the Grundy Library)

Joseph R. Grundy, 1863–1961

Joseph Grundy was raised in Bristol in a Quaker family with deep roots in Bucks County. His father William moved the family worsted-wool business from Philadelphia to Bristol in the early years of the Bristol Improvement Corporation and grew it into a flourishing mill complex.

As the mill owner, Joseph Grundy became a vociferous proponent of manufacturing as the driving force of the American economy and founded the Pennsylvania Manufacturers' Association in 1909. The PMA became a strong lobby first in the state and later nationally in favor of high protective tariffs and opposition to Progressive Era policies that protected workers. (Despite his political leanings, Grundy was known as a generous employer to his own workers, providing educational opportunities and financial help for workers' families if accidents or illnesses struck them.) The PMA became a large contributor to the campaigns of Republican politicians, and made it clear that it expected the men they supported to return the favor by passing laws that protected manufacturers' interests. Though he kept overseeing the Grundy Mills, Grundy became immersed in state and national politics, and at one point was reputed to be the largest financial donor to political campaigns in the U.S. Many historians believe that Grundy was the behind-the-scenes force that won Warren Harding the Republican nomination for president in 1920.

(Courtesy of the Grundy Library)

Grundy was appointed by Pennsylvania's governor to fill an empty U.S. Senate seat in 1929, but was defeated in the Republican primary the following year. He served on Bristol Borough Council for more than thirty years, and was instrumental in developing the town's water and sewerage system and building most of the public schools. Grundy shared Andrew Carnegie's belief that those who had accumulated wealth had a moral and social responsibility to share it. In this spirit, Grundy founded the borough library in memory of his sister. With no heirs, Grundy willed his house and entire estate to a trust that still supports the town, and keeps the Grundy name alive.[10]

In 1900, Joseph Grundy, the son and nephew of the founders, became the sole proprietor of the company. Despite earning a degree from Swarthmore College, he started his career in the family business at the bottom, as a wool sorter, and spent five years working at all the various types of mill equipment used in wool processing. This experience gave him not only a strong grasp of manufacturing methods, but a camaraderie and mutual respect between the owner and his workers—who numbered over 1,000 by that point, making the Grundy Mill the largest employer in Bucks County—that was becoming less and less common in American industry.

The Grundy mill expanded to four buildings along the Delaware Canal, one of which was seven stories tall. In 1910, the landmark 168-foot clock tower was built. Its four clock faces, 14 feet in diameter, are visible in most of the borough. Boatmen on the canal knew they were nearing the end of their downstream journey when they saw the clock tower in the distance. Many blocks in the town still feature rows of solid, nearly identical houses that were built by the Grundy Company and others as worker housing.[11]

"The Diamond City"

One hundred fifty miles north of Bristol is the northern terminus of what is now the D&L National Heritage Corridor—Wilkes-Barre. Like Bristol, it was first settled in the eighteenth century, and is the oldest town in northeastern Pennsylvania. It has been the county seat of Luzerne County since 1786, though it was not designated a borough until 1806 and did not attain city status until 1871. As the population in the anthracite region grew and became concentrated in urban areas, Wilkes-Barre and its larger neighbor to the northeast, Scranton, developed into not only the area's largest cities, but also its principal political, economic, social, and cultural centers.

(Courtesy of Aaron Heckler)

Though Wilkes-Barre's nickname—the Diamond City—came in part from the wealth of anthracite "black diamonds" underlying it, by 1900 coal was not the city's only economic driver. There were 96 incorporated businesses, manufacturing everything from wire rope (Hazard Wire Rope Company) to locomotives and mine machinery (Vulcan Iron Works) to lace (Wilkes-Barre Lace Manufacturing Co.) and thrown silk yarn (Bamford Brothers and Hess, Goldsmith and Company). Wilkes-Barre also boasted the Stegmaier Brewing Company, the largest brewery in Pennsylvania outside Philadelphia and Pittsburgh, and one of the earliest bottlers of beer. These

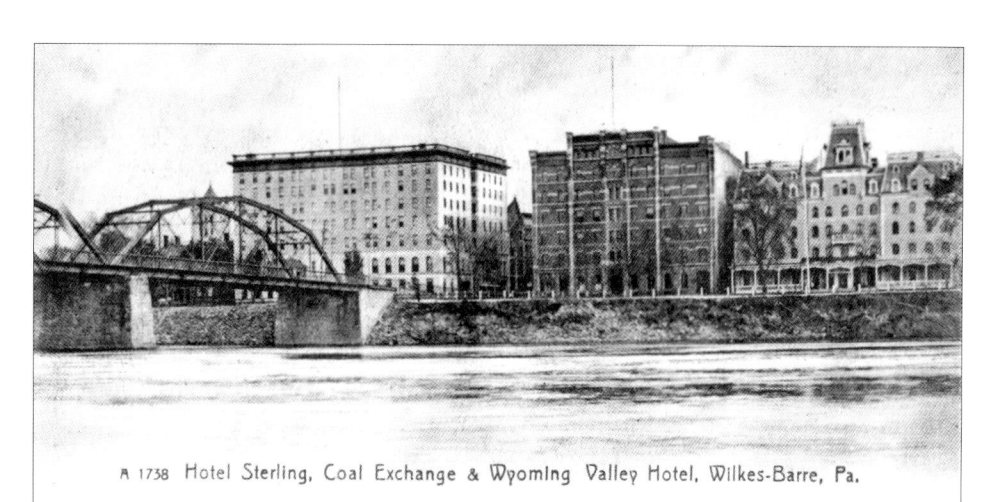

The Market Street bridge across the Susquehanna River in Wilkes-Barre. Two grand hotels flank the prominent Coal Exchange building, which housed a number of businesses in a prime location along the waterfront.
(Courtesy of Ron Bray)

Depiction of an unidentified breaker in the Wilkes-Barre area. (NCM/D&L)

Both postcards predate 1907.

manufacturers employed not only thousands of blue-collar workers, but also a growing number of executives, managers, and clerks who formed a rapidly expanding professional class. Nine banks helped them manage their personal money while financing much of the economic expansion in the city from $2 million in capital assets and over $12 million in deposits.

Six railroads connected Wilkes-Barre to Boston, New York, and Philadelphia, and to northern New York State, the Great Lakes, and Canada, providing transportation for the various commodities produced in the city as well as the most valuable asset—anthracite.

The 163-day strike during 1902 in Pennsylvania's anthracite mines had major national repercussions, and coal shipments during the strike were markedly off from the previous year's record of over 67 million tons from the Wyoming Valley mines. Nevertheless, Wilkes-Barre's commercial enterprises—theaters, hotels, laundries, restaurants, stores that sold everything from furniture to jewelry to oysters, and its 144 licensed liquor dealers—continued to thrive.

The labor peace achieved in 1903 that settled the Great Anthracite Coal Strike allowed Wilkes-Barre and the Wyoming Valley to enjoy growing prosperity and a steady expansion of the anthracite industry until the end of World War I.[12]

The Anthracite Strikes of 1900 and 1902

"It would seem from the conditions of employment and from the general state of industry that, if ever men were justified in asking for an advance [increase] in wages the anthracite miners were in 1900," wrote George O. Virtue, a labor economist, about the anthracite strike of 1900. Writing only months afterward, Virtue sympathetically described the miners' many issues: wages, particularly the despised "sliding scale"; the way each miner's output of coal was weighed; the costs of the black powder that was vital to anthracite mining; company stores and company doctors; and the mine operators' habits of suspending operations to keep the market price of coal high, thus cutting down on the number of days miners actually worked. That, and the obviously rising profits that the coal companies were getting as the nation pulled itself out of the depression of the early 1890s, all combined to let most miners forget their ethnic, language, and religious differences and start to support the United Mine Workers.

Collective action in the anthracite regions had lagged for more than 20 years, since the Workers Benevolent Association had been destroyed in 1875 by a combination of falling coal prices due to the Panic of 1872, and the united front the mine operators presented against the fragmented actions of the miners themselves. The influx of newly immigrated miners from eastern Europe during that period also contributed to Virtue's pessimistic view of the entire anthracite industry, as he wrote in a report for the Bulletin of the U.S. Department of Labor in 1897: "There are too many workmen in the anthracite region trying to make a living by mining coal, and too much capital seeking to make profits."[13]

Virtue's review overlooked several developments. Chief of these were the founding of the United Mine Workers in 1890 and the rising willingness of miners to strike. A multitude of local strikes took place around the anthracite regions in 1899 and early 1900. The leadership of the UMW worked to get everyone who labored in anthracite to understand they had a common cause, and that large-scale collective action was likely to be more effective. A six-week industry-wide strike during September and October of 1900 seemed to vindicate the union's position. The strike yielded pay increases of about 10 percent for most mine workers, but did little to resolve the morass of work rules and pay rates. Still, it was the first successful anthracite strike, and was a *de facto* recognition of the UMW as a union.

The modest success of the 1900 strike bolstered the resolve of many of the mine-company owners never to concede in a strike again. UMW leadership was actually reluctant to call a strike, but by the spring of 1902 it was clear that working conditions and wages were as bad as ever. The union called its members out on May 10. As a result, write Donald Miller and Richard Sharpless in *The Kingdom of Coal*:

> It was not only a work stoppage for current demands—a wage increase of 20 percent and a minimum scale, an eight-hour day, the weighing of coal using a legal ton—but a rebellion against the pernicious past, the lost strikes and broken unions, the dead and maimed, the shattered dreams … Behind the miners' resolve was their conviction that their only hope for an improved existence was a union recognized by the companies.[14]

Pennsylvania's Anthracite Regions

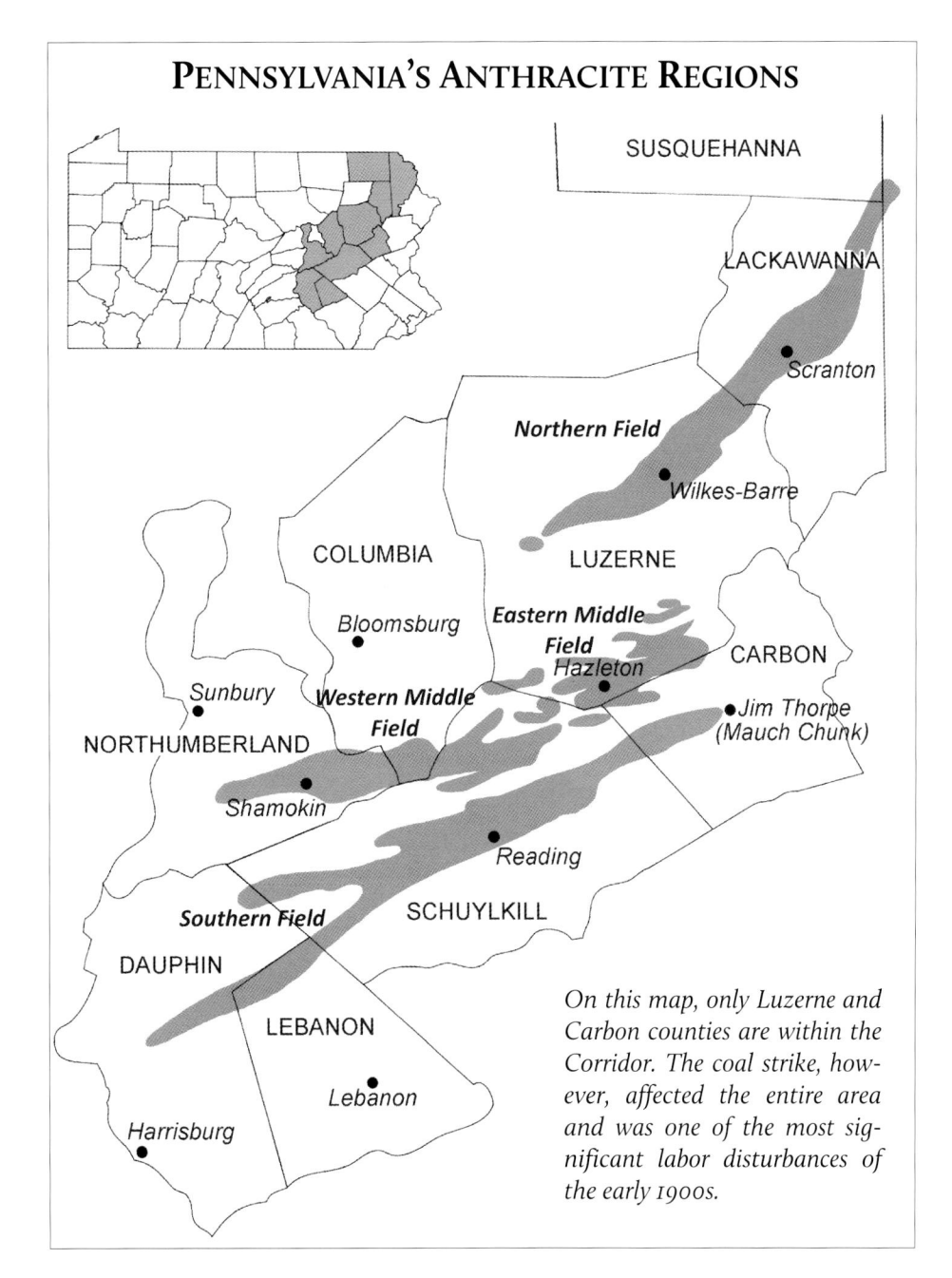

On this map, only Luzerne and Carbon counties are within the Corridor. The coal strike, however, affected the entire area and was one of the most significant labor disturbances of the early 1900s.

With the wages of 140,000 men and boys gone, the local economy dried up. Hundreds of recently arrived Slavic immigrants—mostly bachelors, who were the lower-skilled, lower-paid mine laborers—packed up and either went home or west to find other work. Older children, many barely in their teens, left home to find work in New York or Philadelphia so they could send money home. Others worked long hours, even 13 hours overnight, because it paid 13 cents an hour instead of the usual 10 cents, in the silk mills that were springing up in the coal regions. (Labor scholar Bonnie Stepanoff makes a case that the work of these children helped to prevent the mine owners from starving the miners out.)

The first months of the 1902 strike were peaceful. This changed when the National Guard was called out after a man who was trying to smuggle guns to his brother, a deputy sheriff holed up with two non-union miners in the Shenandoah train station, was murdered by the striking miners massed outside the building. The presence of the National Guard, augmented by 3,000 Coal and Iron Police and 1,000 private detectives, developed a siege mentality among resentful residents that strengthened the resolve of the strikers.

When the strike moved into the summer months, a threat to the conservative leadership of the UMW appeared from the left: radical socialist organizers like the legendary Mother Jones and the anarchist-oriented Workers of the World (the "Wobblies") began holding meetings and recruiting some increasingly desperate strikers.

George F. Baer, president of the Philadelphia and Reading Railroad and its extensive mining operations, wrote a letter on July 17, 1902, in which he claimed that "the rights and interests of the laboring man will be protected and cared for ... by the Christian men to whom God in his infinite wisdom has given the control of the property interests of the country ..." After it became public, the press scathingly pronounced this the "Divine Right" letter, describing it as blasphemy. The widespread negative publicity undercut the position of the owners. Newspaper readers across the country were absorbed by the coal strike.

As the strike crept towards the beginning of the heating season, fears of a coal shortage began to rise. Actually, there was little danger of that happening, as bituminous coal from western Pennsylvania and West Virginia—where the UMW had ruled out a sympathy strike—was reaching the east coast in large quantities. Still, the situation compelled President Theodore Roosevelt to intervene. He called a meeting in the White House of the coal owners, the UMW leadership, the Attorney General, the Secretary of Labor, and himself. The owners' arrogant attitude, in contrast to UMW president John Mitchell's gentlemanly behavior, turned the president's mind in favor of the miners.[16]

Roosevelt proposed an arbitration commission, which the owners rejected. This, coupled

"HURRY UP AND TAKE THE SMALLER ONE, MR. BAER!"
From the *Record-Herald* (Chicago).

Baer's claim of divine support for the coal-mine operators met with widespread outrage among Americans, as seen in this political cartoon from a collection at Ohio State University. The cookie Baer is taking says "Unconditional Surrender."[15]

with widespread newspaper reporting on their meeting with Roosevelt, and the rising public opinion against monopoly ownership in industries, put even more pressure on the owners. The White House leaked a report that Roosevelt was considering sending the Army in to reopen the mines. This got the attention of financier J.P. Morgan, who had vast financial holdings in both mining and the anthracite railroads. Within a day, Morgan forced the mine operators to agree to a commission to examine all sides of the issues.

Morgan was the dominant force in the U.S. economy from 1890 to 1910. During that period, his banking firm directed or oversaw the operations of major corporations such as U.S. Steel, AT&T, and Atlas Cement, as well as 24 railroads. In 1907, he single-handedly pulled the country back from the brink of an economic crash by putting $80 million into bank coffers and securities in four days, ending a panic that threatened to trigger a major depression.

J.P. Morgan

Roosevelt skillfully filled the Anthracite Coal Strike Commission with eminent men, balanced among those who tended to be pro-business, pro-labor, or studiously neutral. Hearings opened in Scranton on November 14; meanwhile, a convention of anthracite miners voted on October 21 to return to work. The commission sat for nearly four months. It heard testimony from 558 people: 240 UMW members, including President John Mitchell, 153 non-union miners, 154 witnesses for the mine operators, and 11 called by the commission. The turning point may have been the 239 other witnesses presented by Clarence Darrow, who led the legal team for the miners. The stories from maimed miners, destitute widows, emaciated breaker boys, children who had never known childhood or education, stories from people of all nationalities, and miners at all work levels, so distressed the commissioners that they called a halt to "The spectacle of horrors." Darrow's impassioned closing speech, delivered for eight hours without notes, tilted the balance.

On March 22, 1903, the commission ruled in favor of a ten-percent wage increase for all anthracite workers, an eight-hour work day, and for weighing the coal loads coming out of the mines to be done by men paid by the miners. It did not, however, recognize the union, and was just the first small step in improving the lives of miners and mine laborers and their families.[17] On the other hand, public outrage about the exploitation of children quickly resulted in laws requiring school attendance and limiting child labor.

Child Labor and Lewis Hine

Pennsylvania passed laws as early as 1885 forbidding children under the age of ten from working in industrial settings such as mines and factories. With little or no enforcement of these laws, children as young as eight or nine were routinely employed ten hours a day, six days a week, at very low wages in textile mills, cigar factories, stores and workshops into the early twentieth century. Progressive-era reformers considered Pennsylvania to be the worst state for child labor, but the greatest outrage was directed at coal mining operations.

Mary Harris "Mother" Jones, 1837–1930

Mary Harris was born in Cork, Ireland, in 1837. Her family fled the potato famine, emigrating to Canada sometime in the early 1850s. Educated in Toronto, she moved to Michigan to teach in a convent school, but soon moved to Chicago and then to Memphis. In 1861, she married George Jones, an organizer for an early guild of iron moulders.

In 1867, George and all four of their children died in a yellow-fever epidemic. Mary moved to Chicago, where she opened a dress-making shop; tragedy struck again when she lost everything in the Great Chicago Fire in 1871. Already painfully aware of the disparities between the wealthy women who bought her dresses and the starving people she saw on the streets, Mary became convinced of the need to do something about the poor wages many working people received. She became involved with some early labor organizations, like the Knights of Labor, and then with the Socialist Party of America and the United Mine Workers. Not an advocate for women's rights or suffrage, "Mother Jones" as she came to call herself, believed that working men deserved wages that let their wives stay home with their children. She was also reputed to say "you don't need to vote to raise hell!"—and raise hell she did.

Mother Jones did not shy away from working directly with men to organize, but also mobilized the wives and children in the labor movement. During the 1900 anthracite strike, she led a large group of women and children in a march from McAdoo to Coaldale in support of the miners, an event commemorated by a PHMC marker along Route 209 in Schuylkill County. Though the UMW leadership did not exactly welcome her brand of radical speech-making during the 1902 strike, she made her presence known and was called "the most dangerous woman in America" by a Pennsylvania legislator for her success in organizing mine workers and their families against the mine owners.

(Library of Congress)

In 1903, she led a group of mill-worker children, who struck in Philadelphia, on a march across New Jersey to attempt to meet President Theodore Roosevelt at his summer home in Oyster Bay, Long Island. Though she failed to get in, and her little band of children ended up back in the mills, Mother Jones had made Americans aware of the plight of child workers.

In her black dress and flowered hat, Mother Jones continued to agitate for workers rights around the nation. Well into her 70s she was imprisoned twice for her activities, and remained outspoken until her death in Maryland in 1930 at the age of 93.

Breaker boys, bare-handed or with hands wrapped in rags, picked unwanted stone and other debris out of the anthracite streaming under their feet in coal breakers, surrounded by massive, loud, and dangerous machinery. Door boys* and mule drivers spent their days in the mines, sharing the dark, cold danger with miners for a fraction of the pay. Families desperate for another wage earner often lied about the ages of the boys they sent to work in the collieries. Child-labor reformer Lewis Hine's photographs of child workers spurred public support of stricter laws, but although Pennsylvania had enacted a compulsory education law as early as 1895, it was not until 1916 that most children under 16 attended school. Changes in breaker technology made breaker boys obsolete by the 1920s.

These photos are from the Lewis Hine collection at the Library of Congress. A door boy is featured on the left. He spent most of the day in complete darkness; this was a job that could be performed by a blind man or boy.

Lewis Hine provided a note about the photo below:

"Holding the door open while a trip goes through. Willie Bryden, a nipper, 164 Center St. [Pittston]. A lonely job. Waiting all alone in the dark for a trip to come through. It was so damp that Willie said he had to be doctoring all the time for his cough. A short distance from here, the gas was pouring into the mine so rapidly that it made a great torch when the foreman lit it. Willie had been working here for four months, 500 feet down the shaft, and a quarter of a mile underground from there. Walls have been whitewashed to make it lighter. January 16th, I found Willie at home sick, His mother admitted that he is only 13 yrs old; will be 14 next July. Said that 4 mos. ago the mine boss told the father to take Willie to work, and that they obtained the certificate from Squire Barrett. (The only thing the Squire could do was to make Willie out to be 16 yrs old.) Willie's father and brother are miners and the home is that of a frugal German family."

---

* The job of a door boy was to control air flow into the working areas of the mines by keeping doors closed, opening them only when mule drivers with mine cars needed to come through.

On the right is a Lewis Hine photo of a group of breaker boys working in extremely dusty conditions.

Below are a posed stereoview of boys in unusually clean clothes picking slate at an unidentified breaker, and a pre-1907 postcard showing breaker boys working at a colliery in Wilkes-Barre.

(Library of Congress and NCM/D&L)

Andrewsville, near Lansford, was a "patch town" where several generations of miners lived in very simple conditions near the mine where they worked. They had space behind their homes for a small plot for growing food. A coal breaker is at the foot of the hill. (NCM/D&L)

Chapter Twelve

1910 TO 1920
LABOR WAR AND WORLD WAR

*"It is a fearful thing to lead this great peaceful people into war, into the most terrible and disastrous of all wars, civilization itself seeming to be in the balance. ...
It will involve the immediate full equipment of the Navy in all respects ..."*

Woodrow Wilson, April 2, 1917

South Bethlehem had seen little in the way of labor unrest, save for a brief, failed strike in 1883, when the Amalgamated Iron Workers tried to organize Bethlehem Iron. When he took over the company in 1904, Charles Schwab made it clear that though he would hear workers' differences and grievances and treat his workers with what, by his lights, was fairness, he would not tolerate unions. And in fact, the 104-day strike that came in 1910 did not start out as an attempt to organize Steel workers. Instead, its roots were the "bonus" system that Schwab adopted based on the work of efficiency expert Frederick W. Taylor, whose studies were designed to boost productivity and reverse the severe downturn in industrial output that followed the Panic of 1907.

The Steel Strike of 1910

The chief complaint of Bethlehem Steel workers was the mandatory overtime that Schwab instituted along with an end to time-and-a-half pay for work outside the normal 12-hour workday. With the company struggling in 1907–1908, Schwab set wages so low that the only way the workers could earn the same amount per week that they had before the Panic was to work for the "bonus." But by 1910, the economy had recovered and the Steel was making record profits. Still, pay was not raised, and not working overtime was not feasible for most of the men. Refusing overtime hours (much of which was Sunday work, which meant a seven-day work week) usually meant being fired.

The trouble started in late January, 1910. A machinist refused to work overtime and was fired. His fellow workers went to the plant superintendent and asked that the man be reinstated, and that Sunday work and overtime be eliminated or returned to time-and-a-half pay. These men were in turn fired. Within days, all the machinists had struck. Schwab then agreed to meet with a delegation of the machinists, but presented them with several reasons why it was impossible to meet their demands. Two days later, the men met with C.A. Buck, the general superintendent of the plant, and told him they were prepared to drop the overtime demand in return for reinstatement of all the strikers. Buck responded by stating that the company would decide who would work there. "In other words," Buck said, "the company wants it understood that the Bethlehem Steel Company is running this plant."

Insulted and outraged, the workers now officially voted not only to strike, but to bring in a union and organize. This strike, however, did not attract the attention of radical labor organizers. In fact, the few union leaders who arrived in Bethlehem quickly saw that the Bethlehem strikers could not sustain a long walkout, and urged them to go back to work and start organizing. Still, the next weeks saw more workers strike, including more than 500 unskilled foreign-born laborers.

Notorious Scabs

Bethlehem, Pa.
March, 1910

Men who deserted their friends and fellow-workers during the great campaign in the year 1910 against the Bethlehem Steel Company for living conditions and against Industrial Slavery.

What Benedict Arnold was to his country so may these men be known as traitors to the cause of labor the world over.

In order that the future may BRAND these men as they properly should be BRANDED, the striking employes of the Bethlehem Steel Company hereby give to the public their names to be inscribed upon the memory of future generations.

NOTE: The names of other traitors to labor's cause will be added and published as soon as reported and found scabbing. Save this circular for future reference.

A four-page circular, "Notorious Scabs," was issued by the strikers during the 1910 Bethlehem Steel strike. Along with the anti-Schwab cartoons shown here, it listed the names and trades of 100 men who either abandoned the strike, or took over jobs that had been held by the strikers. The wide range of ethnic backgrounds of the Steel's workers are evident in the names in the list.

(NCM/D&L)

Mounted police on patrol crossing the old New Street bridge during the strike. (NCM/D&L)

State police lined up for inspection in front of Bethlehem Steel's five-story main office building on Third Street. March 13, 1910. (Private collection)

In late February, the strikers held a five-hour parade through South Bethlehem. Though the event was peaceful, as the crowd was dispersing someone threw a rock though Buck's office window. The next day, a group assembled at one of the plant gates and assaulted both workers going in and the company police guarding the gates.

At this point Northampton County Sheriff Robert Person, urged on by Buck and Schwab's right hand man, Eugene Grace, called in the Pennsylvania State Police. From the minute they arrived, the horse-mounted police behaved belligerently, randomly attacking and arresting men on the streets of South Bethlehem. The violence culminated in a trooper riding his horse up the steps of a hotel and firing several shots into the crowd in the bar room. One man, a Hungarian immigrant, was killed, and several more wounded. Fearing retaliation by the strikers, which never came, more state troopers (then known as "the Cossacks" because of their brutality toward striking miners) were ordered to South Bethlehem.

Though these actions united the residents and small businessmen of South Bethlehem behind the strikers for a time, in late March Schwab hit upon the strategy of threatening to shut down the entire plant unless the community pledged its support for the steel company. Did the community want this to happen, when of the 6,500 workers at his plant only 500 were on strike? Shocked and cowed by the threat to their own livelihoods, business and civic leaders complied, and soon the strikers had nowhere to meet except in an open field. Strikers who left, hoping to find work elsewhere, found they had been blacklisted; their attempts to cripple the company by persuading the United States and foreign governments to void the large contracts they had with it failed.

On May 10, the strikers acceded to Schwab's conditions, and returned to work. Despite a far more radical strike in 1919, it was not until a bloody, protracted strike and intervention by the federal government in 1940 that Bethlehem Steel was unionized.[1]

Charles M. Schwab, 1862–1939

Charles Michael Schwab was born in in Cambria County, Pennsylvania, during the Civil War, and raised in Loretto. Despite his family's modest circumstances, he had two years of schooling at St. Francis College, and in 1879 he moved to western Pennsylvania in search of work. After a brief stint in a grocery store, Schwab went to work as a laborer at the Edgar Thomson Iron and Steel Works in Braddock. He soon attracted the notice of the legendary Bill Jones, the general superintendent of the works, and Jones's boss, Andrew Carnegie. Schwab biographer Kenneth Warren, a professor at the University of Oxford, writes that Jones's ability to communicate his own intense, high-energy competitive drive to his workers to get maximum effort and production from his workers made a deep and lasting impression on the young Schwab, whose own drive and proven ability to get along well with working men led Carnegie to make him general superintendent of the Homestead works at the age of 25. When Jones was killed in an industrial accident in 1889, Schwab was named to succeed his mentor as general superintendent of the larger Edgar Thompson Works.

Schwab was a tireless and constant investigator of improvements to the processes of making steel and steel products, though the owners, Carnegie and Henry Clay Frick, were often less enthusiastic. Schwab's insistence on technical advances, often at the cost of wages, jobs, and sometimes even lives, led to labor troubles. Though often compassionate to an individual's plights, Schwab set his face hard against collective actions, and held fast to the principle that increased efficiency was best in the long run for the business, and thus for its workers.

In 1897, Schwab was named president of Carnegie Steel. Only four years later he secretly negotiated the sale of the company to J.P. Morgan, who combined the various Carnegie works and others into United States Steel. Schwab was named president of the new firm, but in a convoluted and ethically questionable process he also obtained a controlling interest in Bethlehem Shipbuilding and Steel Company. This dual role in competing steel companies could not last, and Schwab resigned as president of U.S. Steel in August 1903, though he remained a director for another year. Now free to devote all his time and energy to Bethlehem, Schwab himself wrote, "Then I really went to work … I put everything I had into Bethlehem." [2]

After World War I, Schwab's personal fortune is estimated to have been as much as $40 million, the equivalent of $300 million today. However, the Great Depression wiped out what little was left after a profligate life that included three sumptuous houses, luxurious private train cars, lavish parties, multiple mistresses (by one of whom he had his only child, a daughter), and heavy gambling during frequent trips to Monte Carlo.

Fifteen hundred mourners filled St. Patrick's Cathedral for his funeral in September 1939. He was eventually buried alongside his parents in a small, dignified family mausoleum in Cresson, Pennsylvania. After his death, Schwab's remaining effects were sold, but the amounts could not begin to cover his outstanding debts. In 1942, Immergrun was sold to his alma mater, St. Francis College, for $32,500, and since then has been a convent and retreat center.

A 1903 portrait of Schwab hangs in the National Portrait Gallery, but with the demise of Bethlehem Steel, the portrait and the family burial vault are his only memorials.

Bethlehem Steel modernized its blast furnaces starting in 1905, and in 1910 moved the main tracks of the Lehigh Valley Railroad, which ran through the center of the plant, to the bank of the Lehigh River. This circa 1910 view from Nisky Hill, across the river, shows the improvements, with the growing town of South Bethlehem behind the plant. (NCM/D&L)

Winning the War with Bethlehem Steel

The outbreak of World War I in August of 1914 was a windfall for Bethlehem Steel. Schwab, already regarded as an industrial genius, looked prescient in having acquired the Fore River Shipbuilding Company in Quincy, Massachusetts, in 1913. Possessing the largest-capacity forging plant in America, and already playing a major role as an international supplier of military hardware, the company was in a unique position to fill orders from the warring powers.

Bethlehem Steel was the first firm in the United States to receive orders for war materials from the allied powers of Britain and France: by December of 1914, Bethlehem had received over $50 million in ordnance orders from these nations, and a total order of $135 million for items such as shells and submarines. This bonanza was particularly welcome; Bethlehem had not paid dividends since 1906 because Schwab had a policy of reinvesting profits in expanding the production capacity of the Bethlehem plant. The

Schwab confers with British Chancellor of the Exchequer and future Prime Minister David Lloyd George in this photograph that may have been taken in 1913. With war looming, the British Admiralty secretly sent for Schwab and the company's top ordnance expert, Archibald Johnston, to negotiate for a large armaments order and twenty submarines. (President Wilson blocked the submarines for fear of damaging U.S. neutrality, so Schwab had them built in Canada with Bethlehem Steel parts.) The company made enormous profits before the U.S. entered the war in 1917, and much more by supplying war materials to the American military when the U.S. declared war on Germany in April, 1917. (NCM/D&L)

influx of war orders also spurred a dramatic rise in the price of Bethlehem's common stock from a level of $30 per share in 1913 to $600 per share in January of 1915, to an eventual peak at $700 in 1916.

To fill the flood of British and French orders, Bethlehem began to expand its facilities at a rapid pace. In 1913 the total workforce of the plant was 9,000 men. By the end of 1914 this had grown to 24,567, of whom more than 2,000 were employed on plant construction alone. Over $25 million was spent on expansion, including a large, four-story addition to No. 2 Machine Shop, which was the company's primary heavy-ordnance finishing facility.

No. 2 Machine Shop. (NCM/D&L)

By the end of 1914 the products of Bethlehem's forging operation were being used in many other parts of the plant to produce military products. No. 3 Machine Shop specialized in manufacturing field artillery caissons, and No. 4 Machine Shop became Bethlehem's primary producer of field guns. Other shops combined to turn out more than 2,000 shells per hour. They were filled with high explosives by a workforce of primarily women at new loading facilities at the company's Redington proving grounds, almost two miles downriver of the main plant.

By 1915 Bethlehem Steel was filling munitions orders from Britain and France, which created thousands of new, though hazardous, jobs. Most were filled by women, who also took over traditionally male work as men enlisted in the war effort in 1917. Women learned skills such as welding fins on mortar shells and operating lathes. The women standing in front of one of the Bethlehem Steel facilities at Redington in 1915 worked at loading shells with explosives. The buildings where this was done had walls designed to blow out in case of an accident. When the troops came home from Europe in 1919, returning soldiers and sailors reclaimed their jobs, shutting women out of these positions until the Second World War. (NCM/D&L)

No. 4 Machine Shop was severely damaged by fire on November 10, 1915, shortly after having been refurbished and new equipment installed to support the armaments business—at a cost of over $1 million. The shop employed 1,500 men in two shifts. Though there was speculation at the time that the fire was sabotage, the company quickly determined that the cause was short-circuiting electrical wires that ignited oil used to lubricate small boring machines. Rebuilding started immediately.

(NCM/D&L)

By the beginning of 1915, Bethlehem had more than $300 million in orders for military products on its books. Expansion of military production was rapid. Between March and August of 1914, Bethlehem manufactured a combined total of 250 gun mounts, caissons, and artillery tubes; between August and December of 1914 it produced over 5,830 completely finished field guns and caisson sets. Shell production rose from 18,620 during the period March to August to a staggering total of 12,792,963 during the next six months. Despite a disastrous fire that destroyed No. 4 Machine Shop in 1915, temporarily decreasing military production, the output of war materials from the Bethlehem plant became ever greater during 1915 and 1916. When the United States entered World War I in April 1917, the importance of Bethlehem Steel as a military contractor rose to new heights.

By 1919 the plant had produced 60 percent of the finished guns ordered by the United States, 65 percent of all American gun forgings ordered, and 40 percent of this nation's artillery shell orders. Its forges and machine shops supplied the French armed forces with semi-finished gun tubes for more than 21,000 field pieces. For Britain and France combined it supplied 65 million pounds of forged military products, 70 million pounds of armor plate, an incredible total of 100 million pounds of steel for shells, and over 20 million rounds of artillery ammunition.

During the years when the United States was engaged in the war, between April of 1917 and the Armistice in November of 1918, Bethlehem produced more than 65 percent of the total number of finished artillery pieces that were manufactured by all of the allied nations and spent more than $102 million building new facilities at the Bethlehem plant.

By 1918 over 31,000 workers were employed at the Bethlehem plant; as many as 5,000 were women, mostly as assembly-line workers in the shell shops. Bethlehem Steel was one of the first corporations in the area to employ women for jobs traditionally held by men, but when the war ended, so did the women's jobs, as did jobs held by men who were not residents of Bethlehem.

Building Battleships, Expanding the Business

The Bethlehem Steel Corporation invested heavily in improvements and additions during WWI, building a large new forging complex in the Saucon Division of the Bethlehem plant to supply the U.S. Navy with 16-inch, 45-caliber guns for its new battleships. The Navy needed new ships, and was planning the largest, fastest, and most heavily armed fleet in the world. Again in World War II Bethlehem built major additions to furnish armaments for the Allies. During the second war, the scarcity of labor was again so great that many women were hired to work in the plant to replace men who had joined the armed services.

The corporation earned enormous profits during World War I. From 1916 to 1922 it invested in acquiring many of the remaining independent steel producers in the United States, making Bethlehem a well-rounded competitor to the United States Steel Corporation in most areas of commercial steel production. It needed significantly more office space to manage the increasing volume of business, and during 1915 and 1916 the Steel General Office building was more than doubled in size. When armaments production was no longer a priority, Eugene Grace, Bethlehem's president since 1916, turned to updating facilities and producing the Grey beam and other commercial products.

With day-to-day oversight of Bethlehem Steel now in the hands of Grace, Schwab and his wife, Emma Eurana, were able to spend more time at Riverside, their palatial home in upper Manhattan. The house, modeled on French chateaux, cost over $5 million when it was completed in 1905. Its 75 rooms, many decorated with interiors taken whole from grand European houses, included a three-story entrance hall with a ceiling-height stained-glass window and the largest and finest pipe organ in New York, and a two-story art gallery full of masterpieces by artists such as Titian, Hals, and Corot.

Architect Maurice Hébert of New York designed Riverside, Schwab's mansion overlooking the Hudson River on Riverside Drive. (NCM/D&L)

Despite an annual income of $250,000 as chairman emeritus of Bethlehem Steel, Schwab could no longer afford the taxes and upkeep of Riverside and he tried to get Fiorello LaGuardia to buy it as a residence for New York City's mayors. LaGuardia thought it too opulent, and the $4 million price tag too high. Ultimately Schwab lost the house and it was demolished. In his last months Schwab, widowed and alone in a "small" Park Avenue apartment, sold his last house, the palatial Immergrun in Loretto, and its contents.

Shells, Grenades, Barbed Wire, and Silk

Other industries in the region also took up production of war materials. Traylor Engineering, established in 1902 as a manufacturer of heavy-duty equipment for the worldwide mining industry, moved into a large facility in Allentown in 1905. The panic of 1907 forced it into bankruptcy, and owner Samuel Traylor had to liquidate all assets, including personal ones, in order to recover.

Traylor Engineering and Manufacturing Company in Allentown expanded its facilities and brought in new equipment in order to make shells for the British government. After the Emergency Fleet Corporation was created, Traylor began production of fifty 500 hp marine engines. (NCM/D&L)

The struggle to become free of debt made him anxious to diversify. When war in Europe began in 1914, he shut down the plant temporarily to re-tool it for making munitions. He sailed to Britain in late 1914, where he won a contract to make one million 18-pound shells to be delivered by January 31, 1915. This and subsequent orders for a variety of munitions, including naval mines, saved the company from the collapse that would have come after the loss of its other non-military overseas contracts. In 1916 the company made $10 million just on its munitions business. By 1917, Traylor's business enterprises, under the name Traylor Engineering and Manufacturing Company, were booming.[3]

Also in Allentown, the American Steel and Wire Company produced most of the barbed wire used by the Allies in the battlefields above the trenches in Europe. As early as 1915, it was a major producer for the Allies, operating 24 hours a day producing more than 100,000 tons of barbed wire, staples, and nails to ship to the Western Front in France and Belgium. The mill went from employing 300 men in 1886 to 1,200 in 1917, working 10- and 12-hour shifts. During World War I, the plant operated 24 hours a day producing barbed wire for the war effort.

Steel ingots were sent by rail from Pittsburgh to Allentown, where they were melted and rolled into wire that was transformed into barbed wire, galvanized wire, and nails of all shapes and sizes. The wire mill was like a small city with its own police force, fire department, and medical staff. A full-time staff of telegraph operators handled the stream of orders from around the world.[4]

Postcard view of the American Steel and Wire Company's plant in Allentown, where barbed wire was made to be sent to the killing fields in Europe. Industries such as this required a large labor force. Allentown was the first city in Pennsylvania to have a War Labor Board, charged with providing skilled and unskilled labor for the essential war industries.

(Private collection)

Explosives maker Pennsylvania Trojan Powder Company in South Whitehall produced grenades and mortars, and was the only maker of hand grenades in the U.S. in WWI. The company first developed a new kind of explosive powder that could be used in place of TNT, which was successfully tested at the company's rural facility near the Jordan Creek northwest of Allentown. The same type of powder was also used in short-range trench mortar shells, which were adopted by the U.S. Army. By the Armistice in November 1918, the company's plants in Pennsylvania and California were producing 50 million tons of the powder annually.[5]

The many silk mills in the Corridor played little role in World War I, other than promoting the purchase of war bonds by employees. In 1919, after the Armistice and while every community was still coming to terms with the war, the Phoenix Silk Company, operators of the Adelaide mills, won the coveted government contract to make multi-colored ribbon for the Victory Medal. This was a medal presented to American and Allied troops who had served in the army during 1914 through 1918. While different nations had slightly different designs on the medal, all used the same ribbon. Adelaide was awarded the contract not because its ribbon was the least expensive, but because the quality and design of its sample surpassed all others. The Adelaide Ribbon Mill in Allentown made 700,000 yards of this rainbow-hued ribbon, with colors in the warp—eighteen in all—for all the colors in the flags of the allied nations. The contract kept twenty looms busy for six months.[6]

Built Like a Mack Truck

In need of an infusion of capital in order to expand their operations—which already employed 700 men in Allentown—and meet the increasing competition, in 1911 the Mack brothers agreed to join the International Motor Company, a Wall Street creation that combined the technical and engineering expertise, the building facilities, and patents of three truck companies—Mack, Hewitt and Saurer. Though this somewhat convoluted new corporation, which sold trucks under all three company names, ultimately saved Mack Trucks, it effectively ended the brothers' management of the company. After 1912, only William Mack remained involved.

The merger brought two brilliant engineers to Mack, Edward R. Hewitt and Alfred Masury. In particular Hewitt, whose guiding principle in design and building was "Keep it simple," built strength, reliability, and ease of servicing and repair into his trucks. By 1916, they introduced the Mack AB and Mack AC models, in part to meet the demands of the British and French for trucks to support their troops at war. The AC retained its chain drive, capable of superior pulling power, and the AB, a heavier model, used the worm, or bevel, gearing system that was becoming the standard. The cabs were steel instead of wood, and had an optional roof, a radical innovation that finally allowed truck drivers to stop sitting like wagon teamsters perched on open wooden benches.[7] The squared-off hoods of the new models soon earned them the iconic nickname of Bull Dog.

In the spring of 1917, the British army ordered quick delivery of 150 ACs to the Western Front in France. Mack Trucks' historian John Montville writes that a story in *Commercial Car Journal* on May 15 that year reported that, in appearance, "these Macks, with their pugnacious front and

View of some of the Mack AC trucks parked along Lawrence Street on city property, waiting to be shipped to Europe in 1918. (Lawrence Street is now MLK Boulevard.) In the left background is part of Traylor Engineering's shops. This site and others where trucks were stored was guarded during the war. International Motors made thousands of trucks for the U.S. Army Corps of Engineers in the war in addition to trucks for the Allies.

(Courtesy of Mack Trucks Historical Museum)

resolute lines, suggest the tenacious quality of the British Bull Dog. In fact, these trucks have been dubbed 'Bull Dog Macks' by the British engineers in charge."[8]

Thousands of 5½-ton Mack AC trucks were shipped to Europe from Allentown, where they were used first by the British Army then by the American Army when the U.S. joined the war in 1918. The iconic bulldog radiator mascot was introduced in 1932.[9]

The demands for fast transportation of people and heavy goods during the war led to recognition that American roads were very deficient. In 1919, five Mack trucks participated in the First Transcontinental Army Convoy that journeyed from Washington, D.C., to San Francisco over the route of the proposed Lincoln Highway. The trip took three months, and made a lasting impression on young Lt. Dwight D. Eisenhower, who recalled it as president when he initiated the Interstate Highway System.

The same year, Mack ACs proved their mettle in two prominent building projects on high California promontories, the Mt. Wilson astronomical observatory, and William Randolph Hearst's "castle" at San Simeon. Both required heavy—and in the case of the Mt. Wilson telescope, delicate—objects to be hauled up steep grades on narrow dirt roads, tasks that only the chain-drive vehicles could accomplish. In only a decade, the Mack name and reputation had spread east and west, far from Allentown.[10]

Shipbuilding Returns to Bristol

When the United States entered World War I in April 1917, Charles Schwab probably knew more about the shipbuilding business than anyone in America. Though Bethlehem Steel was a big presence in the industry, the nation was ill-prepared in every aspect—warships, troop ships, and freighters—for the demands of war on the high seas. There were fewer than 50,000 men working in U.S. shipyards, and the yards in existence had been working at high speed since 1914. Even before Schwab's appointment in April 1918 as Director General of the Emergency Fleet Corporation (EFC), for pay of one dollar a year, the government agency had entered into contracts for construction of 40 ships.

In April 1918, President Wilson appointed Charles Schwab to head the new U.S. Shipping Board's Emergency Fleet Corporation, which was created to address the shortage of merchant vessels available to transport men and supplies to Europe. Bethlehem Steel's president since 1916 had been Eugene Grace, Schwab's protégé.

Schwab was one of many prominent businessmen who agreed to use their professional expertise in the interest of winning the war. They were paid a token $1 a year. These "dollar-a-year" men provided invaluable assistance in the war effort.

(NCM/D&L)

Traylor Engineering of Allentown had been producing munitions for the British and French armies since 1914. The company also became a major manufacturer of large marine engines and boilers, along with other essential parts for ships. In May of 1917, company founder Sam Traylor bought 82 acres of land on the Delaware River south of Bristol and quickly converted the industrial site to a shipyard run by a new company, Traylor Shipbuilding Corporation. The company received an order for ten 3,500-ton wooden freight ships; all ten were completed, though only three before the war ended.[11] However, in October 1918 Schwab decreed that the Emergency Fleet Corporation was ending production of wooden ships and the government would not pay for materials that had already been bought. Traylor Shipbuilding shut down when all ten ships were completed, and was left with a useless yard equipped only to build wooden ships that nobody wanted.

W. Averell Harriman, whose father had made a fortune in western railroads, was better positioned to take advantage of the nation's needs while meeting Schwab's criterion for placing government contracts only with established shipbuilders. He established the Merchant Shipbuilding

This January 1919 photo, part of a very wide panoramic view of Harriman, shows some of the town in the background, and several of the twelve shipways, each with two traveling cranes.

(Courtesy of Grundy Library, Bristol)

Company (MSC) early in 1917 by purchasing the historic John Roach & Sons shipyard in Chester, Pennsylvania, and also establishing a new facility on the banks of the Delaware River just north of Bristol. Merchant Shipbuilding built a dozen shipways with electric cranes and a 1,135-foot fitting-out pier on a 260-acre property there, at a cost of over $12 million.

The wartime powers of the EFC dictated that the facility had to be leased to EFC, but managed by Merchant Shipbuilding. Unfortunately, the location—100 miles from the ocean—required extensive, expensive, and time-consuming dredging to launch ships as large as those Harriman had contracted for. Strikes flared throughout 1917 as the workers' wages were constantly outstripped by wartime inflation.

Harriman acceded to every wage demand, and built a workers' village. Called Harriman, it housed the 3,000 shipyard workers and their families—a total of 15,000 people—and had its own police and fire departments, water and sewerage service, electricity and heating plants, schools, a hospital, and a hotel that served over 12,000 meals per day. Construction of the township was the largest single housing project undertaken by the EFC during World War I. Despite this investment, the delays caused by the dredging and the inexperience of the workers (most experienced shipyard workers were already employed elsewhere) meant that the first ship built in Bristol since the early nineteenth century was not launched until four months after the war ended.

Nevertheless, the backlog of contracts kept both MSC yards busy until early 1921. The problem was that there was no demand for new ships. When the last ship slid off the slips at Harriman, Averell Harriman turned it back to the EFC, which sold the property in 1923. Now no longer home to thousands of people, the wartime township was annexed by the borough of Bristol later that year. The newly enlarged community was about to take flight.[12]

A light bomber, probably a twin-engine, double-tailed Keystone LB-5A. Bristol-based Keystone Aircraft Company built 27 of these planes for the U.S. Army Air Corps in the late 1920s. Note the bat-wing bomber symbol on the aircraft's side.

(NCM/D&L)

An appropriately aerial view of Keystone Aircraft's plant on the former Harriman shipyard in Bristol, taken in the late 1920s. The open space between the building and the Delaware River was the former shipways, and virtually all traces of shipbuilding are gone. However, the adjacent workers' town, Harriman, was annexed by the borough of Bristol in 1921 and many of the former workers' houses are now part of the Harriman Historic District.

(NCM/D&L)

Chapter Thirteen

1920 to 1930
The Roaring Twenties

"Never look back."
Eugene Grace

Many historians consider the 1920s to be the first "modern" decade of the twentieth century. Industrial production rose exponentially as electricity replaced steam power in most manufacturing, and technologies based on assembly-line principles were adopted by more industries. New products—cars, large and small household appliances like washing machines, steam irons and radios, lamps and lighting fixtures, water and plumbing fixtures—all drove a massive, unprecedented increase in spending on consumer goods. A demand for products that would have been unimaginable luxuries only twenty years before in turn created a broad expansion of industries that were "lighter" but no less important than those that drove the economy in the nineteenth century.

But while some industries in the Corridor flourished in the 1920s and laid down foundations that carried them through the rest of the century, those boom years were a bust for both anthracite and merchant pig iron. The decade saw anthracite use decline precipitously from its wartime high, while the last of David Thomas's iron companies, the Crane Iron Company and the Thomas Iron Company, both essentially disappeared.

Ships to Planes

The abandoned Harriman shipyard in Bristol did not remain vacant for long. In 1925, the Huff-Daland Aero Company moved its headquarters to Bristol, and renamed itself the Keystone Aircraft Company. The company quickly expanded to 450 employees as it joined the chase to build aircraft that could make trans-Atlantic flights.

Keystone's Pathfinder plane, named the *American Legion*, showed great promise, but crashed on take-off at its final test flight, carrying a full ocean-crossing load of fuel. The two U.S. Navy pilots aboard were killed. Two weeks later, Charles Lindbergh made his historic flight.[1] Despite that high-profile tragedy, Keystone remained the major producer of biplane bombers for the Army Air Corps, delivering nearly 200 planes by the early 1930s. In 1928, the company purchased the Loening Aeronautical Engineering Corporation, and moved its operations to Bristol, but the following year the newly named Keystone-Loening company was bought by the Wright Corporation, making Keystone a founder-member of Curtiss-Wright, which was created in the middle of 1929.[2]

Keystone continued to make biplanes for the Army until 1932, when the combination of obsolescence and the Great Depression drove it to failure. As we will see in the following chapter, the next tenant of the former Harriman shipyard tract actually combined air and water transportation.

The Dixie Cup Comes to Easton

The early 1920s saw another New York manufacturing firm move west to the wide-open spaces of the Lehigh Valley. In 1919, the Individual Drinking Cup Company, makers of the Health Kup that had recently been re-christened "Dixie," bought seven acres of farmland for $3,000 an acre in Wilson Township outside Easton. In November of that year, it awarded a contract for an 80,000 square-foot, reinforced-concrete factory to a New York construction company. Though on April 1, 1921, as he moved his firm into the vast new building, company president Hugh Moore feared he had foolishly "overbuilt," the following decades proved the wisdom of his gamble.

Health in a "Kup"

Today it seems strange to consider a throw-away paper cup as the icon of a revolution. Yet, two ingenious and energetic brothers-in-law—Lawrence Luellen and Hugh Everett Moore—captured the early twentieth century wave of Progressive Movement health reforms with Luellen's folding paper cup and helped drive the ubiquitous, disease-spreading common water cup or dipper from every classroom, office, workplace, and railroad car in the country.

Luellen, a Boston lawyer, developed a pleated paper cup as well as a vending machine that dispensed the cup, filled with cold water, for a penny. Moore, whose sister Sallie was Luellen's wife, had moved to Boston to attend Harvard, but he dropped out during his second year to help develop the business. Not entirely disinterestedly, Moore also took up the campaign to abolish the common drinking cup. In 1908 Moore and Luellen incorporated the American Water Supply Company of New England to manufacture, install, and service the Luellen vendor.

Hugh Moore, 1940

The same year, Professor of Biology Alvin Davison of Lafayette College in Easton, Pennsylvania, published the results of his study of health and illness in Easton schoolchildren as a result of common-cup contamination. Davison's study proved to be enormously influential, and he testified to the public health-promoting powers of the Luellen vendor several times. The Massachusetts Board of Health sent copies of Davison's paper and endorsement to railroads and school boards throughout the state. But it was Kansas—ironically, the home state of Luellen and the Moores—that was first to pass a law banning the common cup in schools and public areas.

"Once paper cups were recognized as an important tool in disease prevention, state after state began passing laws that forbade the use of a common drinking vessel in public places," wrote Anke Voss-Hubbard in a paper published in 1996.[3] Nevertheless, Moore launched a periodic pam-

Disposable paper cups were first sold in 1908 as a way to prevent the spread of diseases. They were adopted in schools, passenger trains, and public places, replacing the common drinking cup.

The Individual Drinking Cup Company, a pioneer in the business, moved from New York to a large new custom-built plant outside Easton in 1921. Individual-serving Dixie cups for ice cream began to be marketed by the mid-1920s. Custom lids identified the dairy, which had to be a producer of premium ice cream.

The photo shows women making paper cups in the Wilson Borough plant in the 1920s.

(Courtesy of Skillman Library, Lafayette College)

phlet called "The Cup Campaigner" that featured rather ghoulish illustrations demonstrating the risks of disease and death from public cups. Sensitive to the potential charge that he was merely promoting his business, Moore avoided all mention of paper cups, focusing instead on delivering a public-health message. Safeguarding health and preventing the spread of disease became a foundation of the company's marketing message—which continues more than a century later.

Railroads were the first to adopt Moore and Luellen's paper-cup dispensers. The influenza epidemic after World War I boosted sales immensely, though it also aided a rising number of competitors. It was at this point that Moore saw the need for a larger production facility. Once it was firmly established in Easton, the Individual Drinking Cup Company became a marketing and manufacturing dynamo.

First came the "Ice Cream Dixie," the first single-serving, pre-packaged frozen treat. Only makers of high-quality ice cream won the right to serve their product in the Dixie and have their names printed on the lid of the cup, which also bore the Dixie trademark. Advertised widely not only in trade publications but mass-market titles like Good Housekeeping and the Saturday Evening Post, Ice Cream Dixies ads prominently featured children, even though they, presumably, were not the readers. The company directly targeted that audience though the new medium of radio by creating the "Dixie Circus," a wildly popular weekly broadcast. At first, listeners were encouraged to write to the company to request prizes like balloons and pictures of their favorite Circus animal characters. By 1930, this grew into a program of collecting Dixie lids and trading them in for the pictures.

"Ice Cream Dixies earned almost instantaneous consumer acceptance and initiated further diversification of the Dixie line," writes Voss-Hubbard. As the first cups with decorated sidewalls, the little

An advertisement for Ice Cream Dixies published in the Saturday Evening Post, *October 1926.*

(Courtesy of Skillman Library, Lafayette College)

ice-cream cups opened the way for all Dixie cups to have printed decorations on them after 1925. New designs were introduced almost every year, which in turned opened up new ways to advertise and promote Dixie products.[4] This commitment to constant innovation in manufacturing, design, and marketing helped sustain the company through the dark economic days of the Depression.

The End of Iron

World War I was a reprieve for the two remaining merchant pig iron companies in the Corridor, the Crane Iron Works, controlled then by the Empire Steel and Iron Company, and the Thomas Iron Company, which had furnaces in Hokendauqua, Alburtis, and Hellertown. But not long after the Armistice, both companies were hit with the twin troubles of a sudden drop in demand for pig iron coupled with a crippling financial recession.

The Thomas Iron Company had weathered the series of national financial crises of the 1890s and early 1900s under the dynamic leadership of Benjamin Franklin Fackenthal. His successors, however, despite steadily modernizing its two remaining furnaces in Hokendauqua, could not contend with the post-war depression and the influx of cheaper foreign pig iron. One furnace was abandoned in 1924, and the remaining furnace, named the Mary, was blown out in 1927. After the facilities at Hellertown and Alburtis were shut down after producing their last iron in 1921 the company stock was sold to Drexel and Co. in 1922; Drexel subsequently sold the stock to the Reading Iron Company, which eventually sold the remaining works at Hokendauqua to Bethlehem Steel. In an ignominious end to what had been called the finest iron works in the United States, the Steel scrapped the plant in the 1930s.[5]

In Catasauqua, the Crane Iron Works soldiered on for a time, actually dismantling one of its two furnaces and replacing it in 1920 with a larger, high-efficiency furnace that boosted the plant's capacity to 185,000 tons per year. But the Crane, yoked corporately to the rest of Empire's facilities, came under the control of the Warren Pipe and Foundry Company, which was in turn acquired by Replogle Steel in 1924. Faced with the facts that imported European pig iron, which arrived as ships' ballast, could be sold for less than the production costs of locally made iron, as well as American steelmakers who could dump their excess pig iron cheaply, iron production dwindled to a complete stop in 1930 and the plant was scrapped soon after.[6]

The once-mighty Thomas Iron Company was scrapped in the 1930s. All its furnaces in Hokendauqua (above), Hellertown, and Alburtis had been shut down in the 1920s.

The Crane Iron Company along the abandoned canal in Catasauqua, the first of the Lehigh Valley's anthracite-iron furnaces and one of the last to shut down.

(NCM/D&L)

Steel Grows Stronger

While the original industry that had so fundamentally changed the face of the Corridor was declining, the next generation of iron production was becoming ever larger, both in size and economic significance. The last of the anthracite-fueled furnaces gave way to modern furnaces with greatly increased capacity, operated by a corporation that, unlike the merchant pig producers, made iron into steel from which were made beams, foundry items, rails, specialty steels for a new generation of tools, and much else. Led now by Eugene Grace, Bethlehem Steel took full advantage of the enormous profits made by supplying armaments to several nations and of the new facilities built during wartime expansion, and moved its output in new directions during the 1920s—just as it did twenty years later after World War II.

The money made on war production had allowed Schwab and Grace to acquire additional plants such as Lackawanna Steel in New York State, Cambria Steel in Johnstown, Steelton near Harrisburg, Sparrows Point in Maryland, and coal and iron mines in the United States and internationally. Bethlehem Steel in the 1920s was the second-largest steel company in the country.

Eugene Grace

The plant began to concentrate on making steel and steel products primarily for peace-time uses. Its rolled products were especially important as the nation recovered. The H-beams or wide-flange beams made in the Grey Mill had proven themselves and were in great demand. Architects and engineers employed in designing and erecting skyscrapers and long bridges could order beams in almost any configuration they needed from a catalogue listing the hundreds of dimensions available. Rail production, the first and most important product of the original Bethlehem Iron Company in the 1860s, ended in Bethlehem in the early to mid-1920s, after the company had acquired new steel mills that could produce rails more cost-effectively.[7] The rail mill stood where the Combination Mill was later built.

Products made at the Steel were very diverse, and included many orders from other manufacturers for equipment they needed to rebuild after the war was over. Even with the loss of government orders for armaments and munition, the plant thrived as the nation reverted to a peace-time economy. Eugene Grace had a consuming interest in forging ahead, making new products, being the best. Unfortunately he was autocratic and did not like to share leadership, so did not groom a successor as Schwab had groomed him. His passion for the company saved it during the 1930s, when he kept the plant clean, in good repair, and ready to resume full production as soon as the economy picked up. During the worst of times he kept many of his skilled employees working, even if at reduced hours, so they would remain with the company. The Depression, Grace knew, would not last.

The rail-loading yard at Bethlehem Steel. (NCM/D&L)

Slate

During the late nineteenth and early twentieth centuries the leading slate-producing region in the United States was northern Lehigh and Northampton counties. Diverse slate-based products such as roofing materials, black boards, school slates, decorative and functional building materials, electrical insulation panels, fence posts, and billiard tables were produced in Lehigh and Northampton counties. In 1927–1928 the Allentown Chamber of Commerce boasted that "within a 25-mile radius of Allentown in both Lehigh and Northampton counties two-thirds of all the roofing slate in the United States has been produced. In 1912 this region produced 46% of all slate products consumed in the United States and was second only to Maine in the production of electrical slate." World War I, the Depresssion, and the introduction of cheaper materials such as asphalt shingles forced a decline in the industry, and one by one most of the quarries closed.

The date of the first quarry is not known, but was certainly in the first third of the nineteenth century. Slate was hard to transport to market before the canal and railroads. Welsh immigrants William Roberts and Nelson Labar opened a quarry for roofing slate near Kern's Mill (Slatington) in 1845 and within a few years many more slate quarries were operating. Among them were three large deposits developed in Northampton County by Robert M. Jones. Many of the workers were experienced men brought from Wales, Cornwall, and England by the companies.

The train on the right is hauling 65 cars containing 3,750 squares of roofing slate from the Slatington Quarries on the Lehigh Valley Railroad, about 1895. (NCM/D&L)

As the slate industry grew, branch railroad lines were constructed to move the slate from quarries to junctions with the main railroads.

Below are two images of Parsons Brothers' operation in Pen Argyl in 1924. Blocks of slate were lifted out of the quarry using a cableway. A narrow-gauge rail network then transported the blocks to shanties and mills for processing.

(Courtesy of Slate Belt Heritage Center)

Chapter Thirteen: 1920–1930   175

Top right: men being lowered into the hole in "man boxes" at Parsons Brothers' quarry, Pen Argyl.

(Courtesy of Slate Belt Heritage Center)

Quarry men were often of Cornish or Welsh origin, and were brought to the Slate Belt because of their skills and experience working in slate quarries in Cornwall and Wales.

The photo on the right shows a line of the shanties where highly skilled men split slate into roofing slates.
   Below right is the Crescent Slate Company's roofing-slate quarry near Slatington.   (postcards, NCM/D&L)

Waste product is hard to recycle, and piles of it are still visible in the landscape. Some old quarries have filled with water; others have been converted into sanitary landfills.

The Parsons Bros. Slate Company's 500-ft deep quarry (below) was the deepest in the United States.

(Courtesy of Slate Belt Heritage Center)

Silk Rises

As often happens, while one historically important industry was declining in the Corridor, another one was forging rapidly ahead. We saw in Chapter Nine that silk manufacture began to move into eastern Pennsylvania as early as the 1880s. In 1900, there were 25 silk mills in operation in the Lehigh Valley alone, and the industry was on the brink of exploding in the Corridor and the Commonwealth.

As the silk industry grew, so also did a host of ancillary industries. One of the largest silk dyeing firms, National Silk Dyeing, expanded out of Paterson to Allentown, among several Pennsylvania locations, in 1907. Many smaller manufacturers of silk mill equipment were also established, ranging from reeds and creels to quills, bobbins, shuttles, and even the rabbit fur that held quills snugly in the shuttle.

By 1908, there were 299 silk mills in Pennsylvania, the vast majority of them concentrated in the eastern third of the state. One-third of all American silk workers were Pennsylvanians. The majority of them were female, though many weavers and all loom mechanics were men.

The same year, the Pennsylvania State College opened an Extension School in the Stevens School at 6th and Tilghman streets in Allentown. One of the first and most ambitious vocational schools in the country, the curriculum was designed primarily for men working in the city's many silk mills. Evening classes including textile engineering and design, shop arithmetic and higher math, mechanical drawing and other drafting, engineering

The large plant of the National Silk Dyeing Company is above. Its location is along the Lehigh Canal in Allentown, with the switching yards of the Central Railroad of New Jersey at the top of the photo. Lock 40, under the railroad bridge, was called "the dye house lock." Dyeing and finishing textiles continued until the late 1970s here.

(Private collection)

An advertisement for shuttles in a trade journal. Pavia and other companies made numerous products specifically for the growing silk industry.

(NCM/D&L)

Bobbins holding silk yarn sit atop winding machines at the Read & Lovett silk throwing mill in Weatherley. "Throwing" is the process of washing, winding, and twisting raw silk into yarn that can be woven or knitted.

The capacity of a throwing mill was measured in "spindles," the narrow steel rods visible on the machinery in front that hold bobbins being wound with silk yarn. At 42,000 spindles in 1914, Read & Lovett was one of the largest throwing mills in the United States.

(NCM/D&L)

and electrical engineering were offered. George Hammer, a long-time superintendent at the giant Adelaide Silk Mill, began teaching the textile course in 1917, the same year that the firm began covering the tuition for any of its employees who completed the course.[8]

Despite meager wages and long hours, girls and women flocked to work in the silk mills, since these jobs afforded a measure of financial independence and the first real alternative to domestic work available to working-class females.

The percentage of Pennsylvania's silk workers who were immigrants is not clear, but in the weaving mills of the Lehigh Valley many mill employees were from the border region between Austria and Hungary known as Burgenland (many mill owners regarded Burgenlanders as the best weavers in the world), and others came from eastern Europe.

While it is difficult to make precise comparisons between the dollar value of the output of the silk industry and that of industries such as steel, Pennsylvania labor statistics from the 1910s and 1920s show that the number of people occupied by the silk industry equaled, and sometimes exceeded, that of some of the heavy industries.

In 1913, Pennsylvania surpassed New Jersey (which is to say, Paterson, New Jersey, where the silk industry had predominated since the 1840s) as the largest producer of silk goods in the United States. That same year, the U.S. surpassed France to become the world's largest producer of silk textiles, which essentially meant that eastern Pennsylvania, once the world's center of iron production, was now the world's center of silk making.[9]

The majority of the silk operations in Carbon and Luzerne counties, as well as surrounding Lackawanna, Northumberland, and Schuylkill counties, were throwing mills, the plants where raw silk was prepared for weaving or knitting into finished goods. In 1916, of the 213 American throwsters, as these mills were called, 108 were in Pennsylvania, and of those as many as 84 were in the anthracite coal regions.

Weaving mills were concentrated in Lehigh and Northampton counties, while Bucks County became silk-stocking country. This segregation of manufacturing processes reflected the fact that throwing was low-skill, low-paying work, while workers actually producing silk goods for sale required more skills, and therefore higher pay. By separating the operations geographically, silk manufacturers sought—and usually succeeded in the effort—to keep pay-related labor troubles at bay.

The Silk Magnate: D.G. Dery, 1867–1942

Desiderius George Dery was born in Baja, Austria, and educated at the Vienna Textile Academy. He emigrated to Paterson, New Jersey, in 1887, and only five years later opened his own silk mill there. While keeping the highly successful mill in Paterson, he moved his family to Catasauqua in 1896, and the following year opened his first silk mill in the Lehigh Valley. The following year he doubled the size of his mill as well as the number of his workers, to 400. (This mill building, on Race Street in Catasauqua, is now apartments, and is on the National Register of Historic Places.) During the next ten years, Dery opened 15 more mills, all but one in Pennsylvania, and became the largest individual silk manufacturer in the world. At its height, the Dery mills produced 12.5 million yards of broad silk a year, and employed over 4,000 people.

Dery, like Henry Ford, believed it was important to pay his workers fair wages so that they could, in turn, support the other businesses in the communities where he did business. When the U.S. entered World War I, his large contributions to the Liberty Bond drives helped to make Catasauqua the "Million Dollar Town." Over $1 million was raised from Catasauqua alone in each of the drives. His fortune also allowed Dery to build the largest mansion in Catasauqua, which boasted many such grand houses, and he filled his home with art, a science museum and laboratory, an astronomical observatory, a large library, a ballroom and cocktail lounge tiled in Henry Mercer's Moravian tiles, and a large indoor swimming pool.

At the height of Pennsylvania's silk boom, Dery made a fatal business gamble. In 1923, he made a large purchase of Japanese raw silk, only to see China release a vast quantity of the fiber onto the world market at greatly reduced prices. This destroyed his silk empire, and all his mills closed and were sold to other textile concerns. He managed to hold on to his magnificent house until 1934, when, widowed and in serious financial straits, he moved to a small house across the street, where he continued his scientific studies until his death.[10]

(*D.G. Dery in 1939, courtesy of Helen Dery Woodson and Cameron Smith*)

The 1920s were the golden age of silk in the five counties of the Corridor, and through the decade Pennsylvania kept its place as the world's leading silk-goods producer. In fact, it was a period of speculative growth in the industry in this region, much like a financial bubble, as many investors who had no knowledge of or experience in the textile business poured money into building new mills or purchasing existing ones. The industry directory, Davison's Silk Trade, lists many mills, especially in Allentown, that changed ownerships and names annually as poorly run mills failed to return on their investments fast enough.

The boom of the 1920s ended with the onset of the Great Depression. Nevertheless, the textile industry and its cousin, clothing manufacturing, had established a solid foundation in the region, which persisted despite many ups and downs until the 1980s.[11]

Cement Spreads and Settles In

In 1900 there were ten miles of paved concrete roads in the United States. A decade later, the number was only about 14 times higher—144 miles. By 1920, there were somewhere between 20,000 and 35,000 paved roads and streets.[12] World War I had shown that the nation needed to supplement rail transportation with better roads. It was a good time to be in the cement business, and a number of new companies built kilns and opened quarries alongside the flourishing firms that had been strung along the length of the Jacksonburg formation since the 1890s.

By 1920 Lehigh Cement was the nation's biggest cement company in terms of number of plants, with annual production of more than 12 million barrels of portland cement. The company moved into the South in 1923 by building one of the region's largest cement plants in Birmingham, Alabama. In 1925 Lehigh purchased four plants from cement companies: in Alsen, New York; Union Bridge, Maryland; and Bath and Sandt's Eddy, Pennsylvania. These acquisitions, and that of a Buffalo, New York, facility in 1927 brought the company's empire to 21 plants in ten states. Lehigh Cement had net income of $5.9 million on net sales of $30.5 million in 1926. Business was declining even before the Great Depression, however, for in 1929 the company had net sales of only $19.3 million, and its net income had dropped to $2.7 million.[13]

Easton contractor Henry C. Luttenberger received the contract to pave the circle around Centre Square in Easton in 1928. The slogan "100 Minutes from Broadway," the train time between Easton and Manhattan, was adopted in 1910 by the city's Board of Trade. It was to be used by "live-wire men," or economic-development promoters, when discussing Easton's advantages with people seeking sites for industries in the Easton area.

(Courtesy of Ed Pany and the Atlas Cement Company Memorial Museum)

In 1926, the Lehigh District's prodigious builder of cement plants, Fred B. Franks, built the Sandt's Eddy plant in northern Northampton County, and sold it the following year to Lehigh Portland Cement. (Years earlier, Franks had opened the Krause plant in Martin's Creek, later part of Alpha Portland Cement Company, and the Bath Portland Cement Company.) With the capital from the sale of Sandt's Eddy, he constructed the Keystone Portland Cement mill in Bath, just east of his old Bath plant. Keystone used a "wet" process that Franks had devised to make cement until a major overhaul of the facility in the early years of the twenty-first century. Possibly because of Franks's extensive connections and business history in Northampton County, Keystone quickly won large contracts for the cement as paving bricks were ripped up in the streets of Easton and Bethlehem and replaced with smooth, durable concrete.

This aerial view of Pennsylvania-Dixie No. 4 shows the immense size of the plant, and its proximity to the quarry. (NCM/D&L)

The same year, the Pennsylvania-Dixie Cement Corporation was formed out of the consolidation of three companies that had operated along what is now Rt. 248 between Bath and Nazareth. The Penn-Allen, Dexter, and Pennsylvania mills were renamed Penn-Dixie No. 5, 4, and 6, respectively, and greatly expanded.

The Dexter quarry had cement rock much superior to that of the other two mills, so it was the only one kept in operation. Penn-Dixie built a 2½-mile-long aerial tramway to transport the stone between the quarry and the cement mills. The tramcars held just slightly less than a ton of rock, and moved along at 500 feet per minute. (Some of the concrete abutments that supported the legs of the tramway are still visible along Rt. 248.) Cars driving on the road beneath the tram were sometimes pelted with small stones that blew out of the tramcars, but the road was shielded by an overhead steel mesh cover where the tram reached mill No. 6.[14]

Chapter Thirteen: 1920–1930 ~ 181

Above, paving at the busy intersection of Broad and Main streets in Bethlehem, looking east, in September 1928. The steam-powered apparatus is a paver that mixed concrete at the site. The movie theater (left center) was opened as the Kurtz in 1921, became the Colonial in 1924, and the Boyd in 1934.

Above is a Morris Black truck crossing the Broad Street bridge over the Monocacy Creek, delivering Keystone Cement to a road project in Bethlehem.

Workers are installing concrete pavement at Centre Square, Easton, in the photo on the left.

Note that paving bricks remained in place where trolley tracks had been laid.

(All photos this page courtesy of Ed Pany and the Atlas Cement Company Memorial Museum)

The existing cement mills in the Lehigh District shared in the general expansion of the industry. Likewise, industries that supplied the equipment and technology the cement mills needed grew alongside them. Among them was the Traylor Engineering Company of Allentown, a manufacturer of large crushers and mixers for the cement and ore extraction industries internationally. Chapter Twelve recounts the company's expansion into becoming a major producer of shells and other munitions for the Allies during World War I, starting in 1914. While it returned to making equipment for the mining industry during the 1920s and 1930s, World War II made the company a vital defense contractor once again. In 1959, Traylor Engineering was sold to the Fuller Company and continued manufacturing mining equipment until the early 2000s.

In 1916, concerned with diversifying so he would never again suffer the financial problems he had experienced following the economic downturn of 1907, and impressed with the potential of recently developed equipment that sprayed concrete, Samuel Traylor acquired the domestic and foreign rights to a cement gun. Developed by Carl Akeley, the gun was first demonstrated publicly in 1910 at Madison Square Garden but needed engineering refinements. Traylor opened an office in Allentown almost immediately, and purchased the Cement Gun Company in 1920. He and Bryan C. Collier, Cement Gun Company president, continued to improve the machinery that sprayed Gunite (their proprietary name) or shotcrete, and expand its applications.[15] The credit for the success of the cement gun belongs with Traylor, who continually looked for new applications for sprayed concrete, updated its technology regularly, and insisted on hiring salesmen who were not only persistent, but very knowledgeable about its use and willing to target their sales pitches to the specific need of the client. The company remained in Allentown, and in its last years was known as the Allentown Pneumatic Gun Company.

In Catasauqua, James W. Fuller Jr. and his son "Colonel" James W. Fuller III had begun redirecting their former Lehigh Car, Wheel, and Axle Company from supplying components for train cars to producing cement-industry technology as early as 1910. Col. Fuller purchased the rights to the Kinyon Pump, a device that moved bulk goods like coal and cement. (The inventor, Alonzo Kinyon of Allentown, was a fireman on a Lehigh Valley Railroad locomotive. He devised his pump to relieve himself of the back-breaking job of continually shoveling coal from the tender to the locomotive's fire box. Fuller paid him, and later his widow, a royalty for every pump sold.) The Fuller-Kinyon Pump revolutionized the moving of pulverized materials like coal, zinc, and cement in those industries, and was quickly adopted by the electricity industry as a safe and efficient way to feed fuel to steam generators.

Fuller incorporated The Fuller Company in 1926, and established his business headquarters in the former offices of the Empire Steel and Iron Company in Catasauqua. The company continued to acquire the rights to other machinery and technology for moving cement and ores, but contracted the manufacturing with other companies. Finally, in 1936, the company began fabricating its products itself. It acquired ten acres on the site of the Crane Iron Works, and moved its operations into some of the existing buildings. This marked the beginning of the company's steady expansion into national and international cement industries.[16]

Mack Adopts the Bulldog

The nation's thousands of miles of new-paved concrete roads provided a bonanza for Mack Trucks, which opened the decade by renaming itself, and officially making the bulldog its official mascot. Mack Trucks became the corporate name in 1921, and the International Motor Truck Corporation became the name of the manufacturing subsidiary.

Paved highways, as well as a $2 million campaign by Firestone Tire and Rubber Company to encourage Americans to "Ship by Truck," led to trucks becoming a favored way of moving many of the new smaller manufactured goods that were pouring out of American factories to warehouses and retail stores. In all, this new demand led Mack to release nine individual models in two product lines during the 1920s.

Mack not only boosted the horsepower output of its truck engines to the then-astounding rate of 150 hp, it also introduced innovative Rubber Shock Insulators. These were originally intended only for mounting springs on truck frames, but were quickly adopted to mount engines, transmissions, steering columns and radiators in all types of vehicles, making them "one of most widely applied and licensed automotive refinements in history," according to the company's official centennial history.[17]

Mack also introduced power-assisted brakes and a drive shaft instead of chain drive on some models. But the reliable high-torque, chain-drive models were used in the construction of two Western icons—the Hoover Dam, and William Randolph Hearst's fabulous "Castle" at San Simeon, California. During the 1920s, motorized fire apparatus was being produced in any configuration desired by fire chiefs around the country as they replaced their horse-drawn equipment with Mack vehicles. By the end of the decade, Mack Trucks' annual sales topped $55 million, an increase of $32 million in only eight years, with an average profit margin of 10 percent.[18]

This Seitz Brewery delivery truck sports the bulldog hood ornament, which was designed in 1932 to complement the bulldog name that had been associated with International Motors' trucks since World War I.

Easton's Seitz Brewery had used canal boats in the nineteenth century to deliver barrels of beer to canalside taverns along the Lehigh Navigation.

(Courtesy of Mack Trucks Historical Museum)

Massive quantities of anthracite were still being carried by rail through the Lehigh Gorge when this photo was taken in 1924. The view is looking south, with Mauch Chunk in the distance. Industries were then still using anthracite for their steam boilers, and until widespread electrification in the 1930s most residents of the Northeast and Middle Atlantic states were using anthracite for heating and hot water.

The line running to the right is the Central Railroad of New Jersey's Nesquehoning branch. The line moving off to the left and in front of the tower is the northbound CNJ mainline. The now-abandoned LC&N Coalport loading docks are just visible through the signal tower on the left. By this point, no coal was shipped by canal boat from Mauch Chunk, but was carried by rail to a boat-loading dock at Laurys.

In this transportation corridor, the canal, railroads, and road all ran parallel to each other, through the gorge carved out by the Lehigh River. Today's Route 903 is visible further up the hillside. The coal pockets built into the hillside and the water tank above them were for refueling locomotives.

The Lehigh and New England Railroad was acquired by the Lehigh Coal & Navigation Company in 1904, and began hauling Panther Valley anthracite in 1912. It did not run through the Lehigh Gorge. Its tracks in the 1920s are marked by the thick grey line on the map (see p 211).

The LNE was a major northeastern industrial railroad, hauling not only anthracite but also serving cement companies and slate quarries in the Corridor. It also carried passengers, and in 1926 purchased its first gas-electric combination car for passenger service between the Panther Valley and Tamaqua. Operations on the LNE were finally abandoned in 1960.

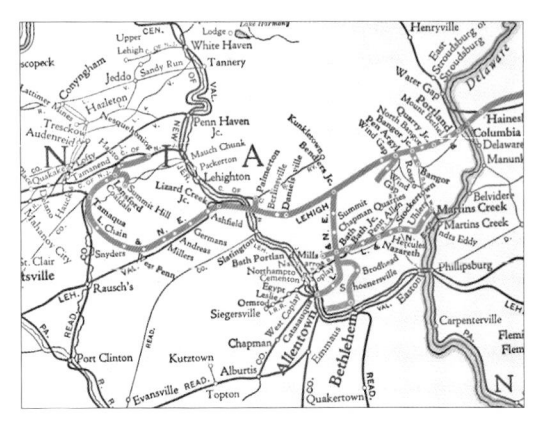

(NCM/D&L; map from the Louis J. G. Buehler Collection)

The Beginning of the End for Anthracite

Anthracite production increased until it reached its all-time peak in 1917; during that year more than 96 million tons of anthracite were mined in Pennsylvania. By this time bituminous coal and coke had superseded anthracite as industrial fuel, and technological advances made it possible to burn oil and natural gas for home heating. The trend toward alternative fuels was given greater impetus when anthracite supplies were cut off by the long strike of 1925. The 1925 walkout was the longest in the industry's history, lasting from September 1, 1925, to February 12, 1926: Markets lost during the strike were never recovered. By 1930 anthracite production had fallen 25 percent from its 1926 level.

The anthracite mining companies were seriously affected by the industry-wide breakup of the railroad and mining combinations, which greatly weakened both the coal-mining and railroad companies. During the 1920s the Lehigh Valley Railroad, the Philadelphia and Reading Railroad,

Delivery of anthracite to the three coal yards on the north bank of the Lehigh River in Easton was complicated, as boats had to cross the slackwater pool behind the dam at the river's mouth. Boat captain Clifford Best told the late canal historian Wouter de Nie that this was done by unhitching the mules and stabling them along the towpath on the south side of the river, then moving the towline to the stern of the boat. Each yard had its own "pin," a metal spike about six inches high, near the towpath, to which the towline was attached. The boat crew would let out the towline as the river's current helped the captain steer toward the opposite shore. After the coal was unloaded, the mules would be re-attached to the towline and pull the boat back to the towpath.

The Hilliard and Dinkey yard above, photographed in 1919, was downstream of the Third Street bridge; the location is now Larry Holmes Drive. (NCM/D&L)

the Delaware and Hudson Railroad, the Pennsylvania Railroad, and the Delaware, Lackawanna and Western Railroad were all forced to divest themselves of their anthracite mining properties. During this period anthracite remained the largest single commodity shipped on these lines, and their mining subsidiaries thus produced much of the traffic on these railroads, but profits were less than if the railroads still owned the anthracite being shipped. All of these railroads had originally been built as anthracite carriers and their route structures were designed for that purpose: the effect of the divestiture was to deprive the railroads of their principal profit centers at a time when they were beginning to face increasing competition from passenger automobiles and long-distance trucking. Due to the broad nature of its charter, the Lehigh Coal and Navigation Company was not forced to divest itself of its railroad subsidiaries, the Nesquehoning and the Panther Creek short lines in the coal regions, the Lehigh and New England Railroad, and the Lehigh and Susquehanna, which was on long-term lease to the Central Railroad of New Jersey. These lines, in particular the Lehigh and New England, remained major profit centers for this company until the late 1950s.[19]

By 1920, the population of the anthracite coal regions had topped one million, and approximately half of those residents were either foreign-born or the children of immigrants. The decline in mining triggered a slower, but equally significant decline in population—10 percent each decade since 1930.[20] (The estimated 2018 population of Luzerne, Lackawanna, Carbon, and Schuylkill counties is 734,523).[21]

Far more dramatic has been the decline in the number of people working in the anthracite industry, which has fallen from a peak of 175,000 in 1920 to approximately 900 now.[22]

Many of the mining jobs were never replaced by newer industries; as a result, many former miners took jobs in the Lehigh Valley at businesses like Bethlehem Steel and Mack Trucks. Tiring of the long commutes, and seeing few opportunities for the future for their children, many coal-region residents reluctantly pulled up stakes and moved away, an exodus that continues to the present time.

Chapter Fourteen

1930 to 1960
Depression, War, Booms, and Busts

"There are many ways of going forward, but only one way of standing still."
Franklin D. Roosevelt

With the exception of the merchant pig iron business, the major manufacturing industries in the five counties of the Heritage Corridor survived the Great Depression, though some came through it in better shape than others. Most persisted through the balance of the twentieth century, though the decline of manufacturing in the United States as a whole began to be perceptible in the Corridor—particularly in the textile and needle trades—by the 1980s. Anthracite mining declined rapidly in the post-war period, and the 1959 Knox Mine disaster, which resulted in extensive mine flooding, ended underground mining in the northern anthracite fields.

Economic growth and development during the 25 years following World War II in the United States as a whole was evident in most areas of the Corridor. Alongside most of the major industries that rose during the nineteenth and early twentieth centuries, companies in a wide variety of fields, such as Air Products and Chemicals, Alpo, Crayola, Just Born, and Nesquehoning's KME (originally Kovach Mobile Equipment) grew rapidly and expanded their markets nationally and even internationally. This diversity of manufacturing broadened the economic base of the region and created a wider spectrum of white- and blue-collar jobs, and softened the effects of the decline and even disappearance of the industries that had been dominant before.

Here's what befell those early industrial giants of the Corridor:

Bethlehem Steel

The Steel's string of successes before, during, and after World War I left it with deep cash resources, and the effects of the Great Depression were mitigated by the U.S. Navy's build-up in the 1930s to prepare for the perceived threat of Japan's fleet. Though major, the company's role in defense was not as dominant during World War II, and the post-war years saw the end of armor-plate production. Nevertheless, the next thirty years were generally very prosperous for the company and its employees.

Bethlehem Steel's combination mill was the most modern rolling mill in the world when it was put into production in 1968, at a time when the company was the largest producer of structural steel in the U.S. It was called the combination mill because it could be switched between making wide-flange and standard structural sections. Computer-assisted tomography developed at the company's Homer Research Labs allowed for taking measurements during the rolling process and made on-the-spot adjustments possible. The combination mill turned out a wide variety of products in various sizes, including Bethlehem's signature "H" beams, in as fast as one minute per piece. However, the slowing demand for structural steel, competition from "mini-mills," and the end of steelmaking at the plant led to the mill's shutdown in March, 1997. (NCM/D&L)

Demand for steel was never higher than in the 1960s, and huge investments were made in improving and expanding the company's facilities in Bethlehem and elsewhere. Structural steel became the firm's core product, with Bethlehem the nation's largest producer. The combination mill that opened in 1968 was the most modern in the world, able to switch between wide-flange and standard structural shapes. Some innovations were too late: the basic oxygen furnace put into production in 1968 was state of the art, but that technology was more than a decade old, and reliant on being close to blast furnaces for fresh molten pig iron. By the late 1960s so-called mini mills that reprocessed steel scrap were starting to encroach on the profits of integrated steel mills like Bethlehem. Mini mills did not need blast furnaces or coke ovens, which meant far lower labor and capital costs, so they could charge less for their steel. Furthermore, the facilities of the mini mills were designed with a far more efficient layout for twentieth-century metal production than older plants such as in Bethlehem. The lighter steel they produced was soon much in demand by automakers responding to the gas shortages of the 1970s with smaller, lighter cars. The replacement of structural steel by precast and pre-stressed concrete for buildings and bridges also put pressure on the Steel's bottom line.

It was the Bethlehem plant's inland, landlocked site that eventually became the biggest impediment to its survival, wrote retired director of strategic planning John Lovis. Other steel mills that Bethlehem owned, located on the Great Lakes or Chesapeake Bay, could get bulk quantities of coke and iron ore delivered directly by water, while the Bethlehem plant's supplies had to be offloaded from ships and carried in by rail. This raised both handling and labor costs. The physical layout and building congestion of the plant made it impossible to move to the continuous-casting method, a process that had significantly reduced costs for newer plants. The upshot of

The last cast of pig iron pours from Blast Furnace "C" into "submarine cars" on the afternoon of November 18, 1995. After the blast furnace shut down, the Basic Oxygen Furnace had no source of hot metal to make steel, so this was the end of the "hot" side of the Bethlehem plant.

Two members of the crew of Blast Furnace "C" on the last shift.

(Both photos: Joe Elliott for NCM/D&L)

Above, an ocean-going ore carrier is unloading its cargo alongside the blast furnaces of U.S. Steel's Fairless Steel works on the Delaware River, which was located a few miles upstream from Bristol. This kind of deep-water access for the delivery of raw materials became essential for economical operation of steel mills. However, steel was made here by the open-hearth method only, which was obsolete when the plant opened, and its working life was less than forty-five years. (NCM/D&L)

all these pressures was the shutdown in November 1995 of the "hot" side of the plant—the blast furnaces, the BOF, electric furnaces, foundries, and the Grey Mill. The combination mill continued in operation for two more years, using blooms produced at Bethlehem's Steelton plant and brought to Bethlehem on specially insulated rail cars, and the coke works shut down in 1998. The entire Bethlehem Steel Corporation filed for bankruptcy in 2003.[1]

An important segment of Bethlehem Steel lives on. The heavy forging that began with battleship armor plate continued through the twentieth century, with huge forgings for turbine and generator shafts for electrical generation and components for nuclear reactors. In 1992, the forge department was spun off into an independent business unit known as BethForge. Like the combination mill, which did not survive, it continues to operate with blooms produced at Steelton. BethForge was sold to WHEMCO Incorporated in 1997, and renamed Lehigh Heavy Forge. It is the sole remaining super-heavy forge in North America.[2]

Cement

The Lehigh Valley's dominance of cement manufacturing decreased rapidly as the industry became established in other regions of the United States. Cement prices in the United States reached a peak of $2.02 a barrel in 1930. Consumption fell from a high of 72 percent of capacity in 1928 to only 46 percent of capacity in 1931, when prices dropped to $1.15 a barrel. The cement industry as a whole lost $25 million that year. Lehigh Cement, the nation's second largest producer, still made a profit in 1931, but it lost nearly $2 million in 1932 and $592,000 in 1933. Subsequent years were profitable, even though net sales fell as low as $9 million in 1935. The company's president, Joseph Young, later told a reporter, "It was only by throwing eight plants overboard that we were able to ride out the storm of the Depression." Lehigh Cement abandoned several plants: one of the two in Mitchell, both Ormrod plants, the ones in West Coplay and Bath, and all three New Castle facilities.[3]

Cement is a cyclical industry, vulnerable to upward and downward trends in national and international economies. For decades the industry has consolidated its smaller plants into large ones with greater capacity and higher efficiency, in the Lehigh Valley and other areas. Even when mills run at top production levels around the clock, they now employ far fewer employees than they did even 50 years ago. Local cement companies have adopted various types of waste as part of the fuel source in their kilns—such as used tires, chemical and medical after-products, and other forms of chemical and hazardous waste. The industry maintains that burning such often-toxic refuse at the extremely high temperatures attained by cement kilns, though controversial, is a safe, economical, even environmentally friendly means of disposing of it.

Four corporations currently produce cement in the Lehigh District, each operating mills and quarries that have been in use for at least the past 90 years. All five cement plants (Lehigh Hansen owns two) are owned by foreign concerns—French, Spanish, and German—and serve domestic and foreign customers.

In 1959 the Fuller Company, maker of equipment and technology for the cement industry, bought Traylor Engineering of Allentown, whose products were mining technology and equipment. The combined companies were subsumed into the Danish company FLSmidth in 1990. FLSmidth had been Fuller's chief international competitor in the industry, and currently maintains offices in the Lehigh Valley. The company still produces "Fuller-Traylor" mining and crushing equipment, perpetuating the name of two major Lehigh Valley industries. A third international company, Bradley Pulverizer, which was founded in 1886 to provide equipment to manufacturers of cement, remains headquartered in Allentown.

Silk

The Great Depression, the development of synthetic textile fibers, and the shift of most of the U.S. textile industry to the southern states all severely damaged the Pennsylvania silk business. World War II completely disrupted the importation of raw silk from Asia. However, silk mills—many combining silk and synthetic production—were still an important regional industry during the 1950s and 1960s. Some weaving and knitting mills remained in business as late as the 1990s, but the sound of the silk loom has long been silent in the Corridor.

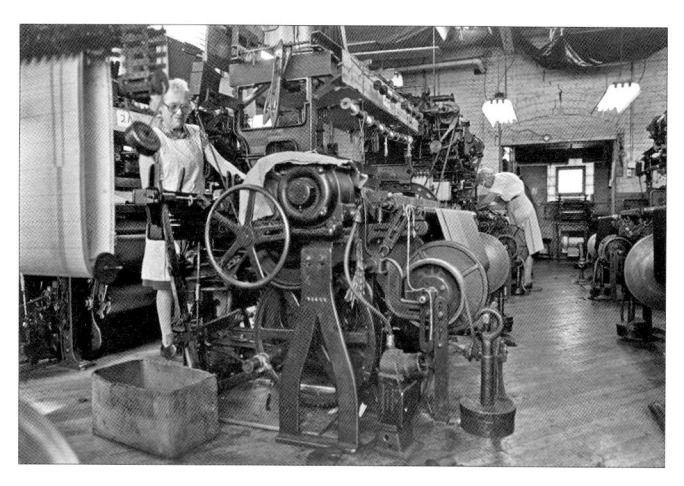

Two weavers work at silk looms in the Catoir mill in Allentown, photographed by Ken Bloom in 1982. When the mill closed in 1989, it was the last silk mill in the Lehigh Valley and the remaining workers, like these women, were all well past conventional retirement age. (NCM/D&L)

Anthracite

During the 1930s anthracite production declined further due to the Great Depression and increasing competition from alternative fuels. A short-lived respite came during World War II, because of the great need for any type of fuel to support defense industries. This temporary boom period came to an end by 1948. During the 1950s the traditional underground anthracite coal-mining industry entered its death throes. In 1954 the Lehigh Coal and Navigation Company ceased to operate its Panther Valley mines, although other companies and later a miners' cooperative leased many of these properties and continued deep mining until 1972.

Anthracite photographer George Harvan captured these miners working in the "Mammoth Vein" in the 1940s, when it was owned by Lehigh Coal and Navigation. Today, almost all underground mining has ended here and throughout the anthracite region. The Mammoth Vein is composed of a very pure form of anthracite, and was the thickest vein found in Pennsylvania's anthracite fields.

(NCM/D&L)

The Nesquehoning Breaker, seen here in about 1930, was built in 1908, replacing one used since 1886. LCN collieries (the collective name for the mines and the breaker where coal from those mines was processed) were numbered 1 through 10 from east to west. Nesquehoning was designated No. 1. This breaker was shut down in 1948 as LCN consolidated operations when demand for anthracite declined after World War II. Coal from the No. 1 mine was then hauled to the company's newest breaker at Lansford. No large breakers are still standing in the Corridor. (NCM/D&L)

Fauzio Brothers Stripping and Hauling Contractors was the largest stripping contractor for the Lehigh Navigation Coal Company's mining operations in the Panther Valley. When it ceased mining in 1954, the Fauzio brothers continued strip mining for two companies that leased the coal properties. In 1960 the brothers' company, Greenwood Stripping Corporation, leased all LCN's anthracite properties and then bought them outright in 1966 for nearly $1.5 million. In 1974, they sold the entire property to Bethlehem Mining Corporation, a subsidiary of Bethlehem Steel, for a reputed $20 million. (NCM/D&L)

During the 1950s many of the coal-mining properties in the Wyoming Valley were consolidated under the control of the Blue Coal Company. Tragedy occurred when miners broke through to the Susquehanna River at a Pennsylvania Coal Company mine, killing twelve miners.[4] The flood caused by the disaster soon ended almost all underground mining in the Wyoming fields. The mines remain flooded and are largely inaccessible today.

By 1960, total annual production of anthracite had declined to under 19 million tons and in 1973 annual production had fallen to 6,746,660 tons. In 2016, 7.6 million tons of anthracite were produced, with over half—4.4 million tons—reclaimed from coal-waste piles. Only ten underground anthracite mines, producing a little over 90 tons, were operating, eight in Schuylkill County and two in Northumberland County. Fifty-three surface mines, which are digging down to former underground mines, produced another 3 million tons. Only 900 people work in the industry.[5]

Though some residents of northeastern Pennsylvania still heat their houses with anthracite, the bulk of production is used in water filtration plants and in metallurgy. Some is exported.

Mack Trucks

The Depression had a devastating effect on Mack. In addition to the drop in demand, light-duty trucks introduced by other manufacturers created competition for Mack's large models. Mack sales dropped 75 percent between 1929 and 1932. But the company fought back by offering smaller trucks and its own diesel engine, and introduced cab-over-engine models. This design, the best way of getting a distribution by weight on the front and rear axles, was necessitated by laws restricting axle loading, gross vehicle weights, and overall lengths.

Despite the Depression, Mack's new line was successful. Since trucks offered the lowest cost per mile to transport manufactured goods, companies in financial distress increasingly turned to them. Furthermore, the urban demand for public transit ensured a strong market for buses, which Mack produced until early 1960.

Mack made this 1941 pumper for the Goodwill Hose Company in Bristol.

(Courtesy of Mack Trucks Historical Museum)

A 1943 Mack NM6, produced in large numbers for the U.S. Army during World War II for moving troops and cargo.

(Courtesy of Mack Trucks Historical Museum)

Mack's leadership of the industry continued in 1938 with the introduction of the Mack Diesel, the first diesel engine made by a truck manufacturer. In 1940 Mack sales hit $44 million on domestic deliveries of 7,754 units, with a net profit of $1.8 million. The company's role in producing vehicles for American and Allied forces during World War II was significant in terms of both numbers and types.[6]

By making heavy-duty trucks, small delivery trucks, dump units, buses, and fire trucks, Mack offered the most comprehensive product line of any truck manufacturer. Although production of Mack's famous fire apparatus ended in 1990, many communities still operate their Mack fire engines.

Long-haul trucking took off with the construction of the interstate highway system in the 1950s and 1960s. However, despite the quality of its products, during the 1950s and 1960s Mack declined steadily. By the time Zenon Hansen took over as chief executive in 1965, the company's prospects of surviving were considered very poor. Hansen had worked in the trucking industry his entire career, and knew what the industry needed. He initiated product-focused changes, whereas previous executives had focused unsuccessfully on finances. He began an aggressive marketing campaign to revive the almost forgotten bulldog emblem, and boosted employee morale through better communications. The company quickly started to turn around in both production and profitability. The large assembly plant in Lower Macungie Township has been a major employer in the Lehigh Valley ever since it opened in 1975 and today is producing the Anthem, Mack's newest heavy-duty highway truck.[7] The Allentown plants closed in 1987. Mack was acquired by AB Volvo in 2000, and the corporate headquarters, opened in 1970, were moved from Allentown to Greensboro, North Carolina, in 2010.

Aviation in Bristol

The aviation industry persisted in Bristol into the space age. Keystone Aircraft Company was purchased by the Wright Corporation, becoming a division of what became Curtiss-Wright. Keystone developed the Patrician, which boasted a top speed of 151 miles per hour, and was the one of the first planes for daily civilian travel. Keystone's long-standing defense contracts sustained it through the earliest years of the Depression, and it delivered the last 39 bi-plane bombers purchased by the Army Air Corps. But the Depression caused a deep decline in U.S. military spending, and Keystone closed in 1932.

A Fleetwings Seabird makes a run on the Delaware River near Bristol in the 1930s. The amphibious monoplane "flying boat" was built of stainless steel, which would not corrode in fresh or salt water and could be spot-welded instead of riveted to save weight. The engine was housed in the nacelle above the wing, which kept the cabin unusually quiet. Of the six Seabirds that were built, at least one is still air-worthy. Housed at the Golden Wings Flying Museum in Blaine, Minnesota, it makes occasional appearances at air shows. (NCM/D&L)

The site on the Delaware was ideal for building airplanes, though, and in 1934 Fleetwings, Inc., moved from Long Island to Bristol to begin building amphibious planes. The Sea Bird, a spot-welded all-stainless-steel aircraft, was a technological success. But its $25,000 price tag was too high for the Depression economy, so only six of the planned 50 were built. In 1943, Kaiser Industries bought the company, which produced components such as tail assemblies for military aircraft. Kaiser-Fleetwings' final signature product was the launch canister for the Echo 1 balloon satellite in 1960. The facility closed the following year, and was demolished for housing.[8]

Railroads

As demand for coal declined across the country, and the anthracite industry collapsed altogether, railroads that served the Northeast fell into dire financial straits. Between 1967 and 1972, six major northeastern railroads—the Lehigh Valley, the Reading, the Erie Lackawanna, the Central

Lehigh Valley Railroad locomotive No. 2102, the John Wilkes, barrels through Lehigh Gorge sometime between 1939 and 1948. A 1916 Baldwin 4-6-2 Pacific-class, the John Wilkes was streamlined in 1938 and made the run between New York City and Pittston, PA. The train was named for an English politician and journalist who supported American independence, and who is also memorialized in the name Wilkes-Barre.

(NCM/D&L)

Railroad of New Jersey, the Penn Central (formed in 1968 by the merger of the Pennsylvania Railroad and the New York Central), and the Lehigh and Hudson River—went bankrupt. To preserve some passenger service, the U.S. government created Amtrak, but more pressing was the question of what to do to save freight lines. Although the tonnage of freight carried by railroads had plummeted, rail transport was still vital to many industries, such as autos and steel, in the Northeast, mid-Atlantic and upper Midwest, as well as serving many major U.S. cities.

When the Penn Central failed, Congress passed the Regional Rail Reorganization ("3R") Act in 1974. It provided emergency funding for the bankrupt lines, and created Consolidated Rail Corporation, or Conrail, as a federally funded private company. Congress also charged the U.S. Railway Association with identifying lines in the bankrupt railroads that could be incorporated into a new, more efficient system. Obsolete, redundant, or irreparable trackage, stations, trestles, and other assets of the old lines were abandoned, and personnel was drastically reduced.

Conrail began operations on April 1, 1976, with a mandate to revitalize rail service in the Northeast and Midwest and to make money doing it. This goal was elusive until the passage of the Staggers Rail Act in 1980, which allowed railroads to set rates for carrying freight and thereby be competitive with truck transportation. By 1983, Conrail was the fourth-largest freight hauler in the country, and was turning modest profits. In 1987 Conrail was returned to the private sector in what was then the largest initial public offering in U.S. history, raising $1.9 billion.

In the spring of 1997, the two remaining Class 1 railroads in the East, Norfolk Southern Corporation (NS) and CSX Corporation (CSX) agreed to acquire Conrail through a joint stock purchase. CSX and NS split most of the company's assets between them. Conrail still exists as a jointly owned subsidiary of each; in addition, Conrail itself still owns three Shared Assets Areas in Detroit, Philadelphia, and northern New Jersey, where the line provides terminal and switching services.[9] Conrail can fairly be said to have saved rail freight in the eastern United States. Today, Norfolk Southern operates more than 21,000 miles of track, including much of the main lines of the former anthracite railroads. It is one of the largest freight haulers in the U.S., the largest transporter of auto parts and complete cars in the eastern U.S., the largest intermodal operator in the east, such as the train-to-truck facility on the site of the former Bethlehem Steel coke works, and a major carrier of bituminous coal. As of 2017, bituminous from eastern coalfields was still its most common commodity freight.[10]

Canals

By the spring of 1931, only twenty boats remained in operation on the Lehigh and Delaware canals, and only 65,000 tons of coal were transported. Most of it was delivered to coal yards in towns like Yardley, Bristol, and Morrisville in southern Bucks County that had no other way to get the coal, as no railroad with connections to the anthracite regions served those communities. Even so, during the 1920s, heavier trucks and improved roads diverted a significant amount of the anthracite from canal boats.

Beginning in June 1923, no anthracite was shipped by boat out of Mauch Chunk because the Navigation was so clogged with coal silt. Instead, it was shipped by rail to a transfer station and dock at Slate Dam at Laury's, and also apparently, according to the waybill pictured here, to Siegfried's (Northampton). By this time, most anthracite was used only for heating.

The onset of the Great Depression convinced LC&N, which had controlled the Delaware Division since leasing it in 1866, to end commercial navigation on the Delaware Canal. On October 18, 1931, the deed to 40 miles of the Delaware Canal between Raubsville and Yardley was ceremonially handed over to Governor Gifford Pinchot, who accepted it on behalf of the Commonwealth of Pennsylvania.[11]

The Lehigh Canal stayed in partial operation for a few more years, hauling some coal to the remaining canal-side coal yards and dredging coal silt from the river and canal and hauling it to Palmerton for fuel for the zinc smelter. The cost of repairs to the waterway from repeated floods in the 1930s burdened the LC&N, and after two severe floods in the spring of 1942, the Lehigh Canal was abandoned.

The waybill above accompanied a load of coal transported in August 1931, close to the end of operations on the Delaware Division.

On the right is Leedom's coal yard in Yardley, displaying an Old Company's Lehigh sign.

(NCM/D&L)

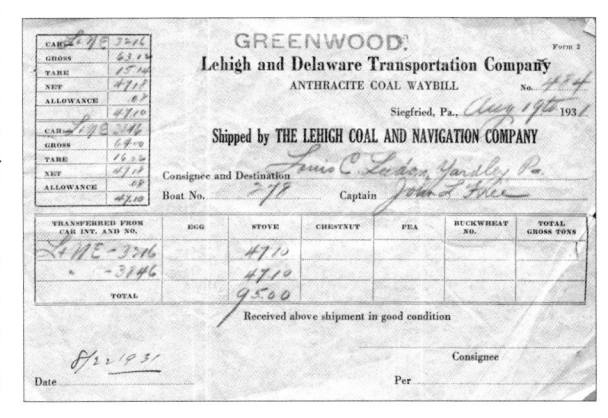

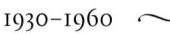

Even before navigation ceased on the Delaware Canal, some LC&N work scows were converted to excursion boats that carried Sunday picnickers. The trips were limited to one level of the canal, as the locks did not operate on Sundays. The level between the entrance lock to the canal at Easton and Lock 22 at Raubsville was popular with groups from Easton. The Young Men's and Young Women's Hebrew Association of Easton paused for this photo during their annual "Camptown" canal-boat trip in August, 1939.

(NCM/D&L).

In 1962, LC&N began selling off its Navigation properties in preparation for returning control of the Lehigh River to Pennsylvania. Hugh Moore, the retired genius of Dixie Cup, donated nearly $20,000 to the City of Easton to purchase the land in Section 8 of the Lehigh Navigation. This totaled 260 acres, from Hopesville in Bethlehem Township to the entrance to Delaware Canal State Park just below the confluence of the Lehigh and Delaware rivers. Moore turned his considerable wealth over to several philanthropic pursuits, including active support for the Marshall Plan and the United Nations, world population control, and conservation of land and wildlife. In making the former canal and its surrounding lands open to all, Moore's vision was to provide "regions of beauty and quiet" for area residents and a park "that could mean as much to us as Fairmount Park does to Philadelphia." Moore's support of the park exceeded $400,000 in his lifetime, and continues to support the park that bears his name through the Hugh Moore Trust.

The 2.5 miles of the canal through Hugh Moore Park were fully restored in 1976. The National Canal Museum has operated a seasonal, mule-drawn canal boat ride in this stretch of canal since then. In 2012, the museum moved to a building along the canal that was already the headquarters of the Delaware & Lehigh National Heritage Corridor; a merger of the two nonprofit organizations, which share a focus on the historic transportation pathways that carried anthracite to market by water and rail, was completed in 2018.

The D&L Trail, on the Lehigh and Delaware Canal towpaths and the former Lehigh Valley Railroad right-of-way, is in 2019 nearing completion of its restoration of the historic connection between Wilkes-Barre and Bristol and all the Corridor communities in between.

198 ~ Geography, Geology, and Genius

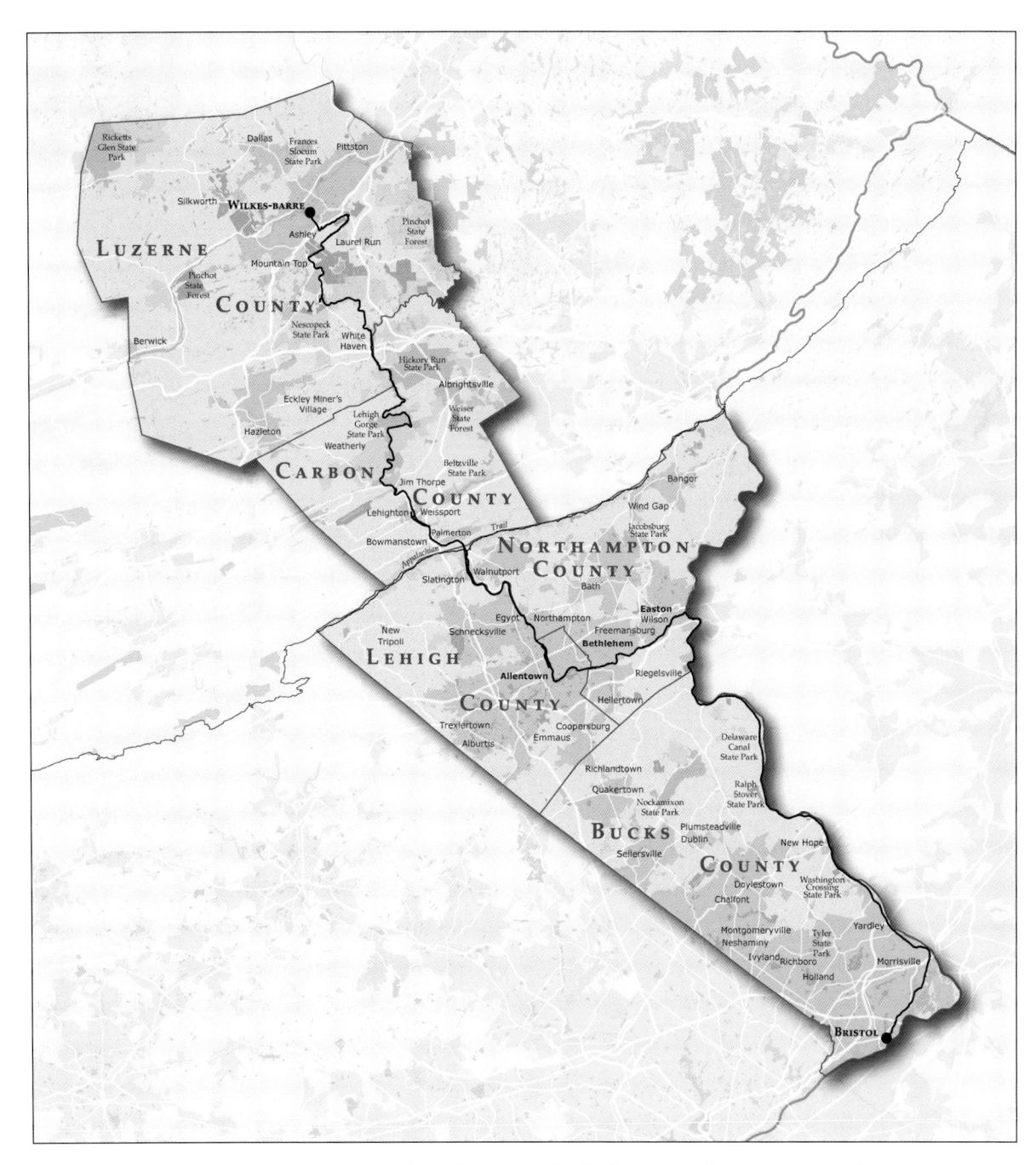

The counties comprising the Delaware and Lehigh National Heritage Corridor.

Epilogue

This work is not meant to be an exercise in recalling the past glories of the region of the Corridor and painting them as the now-vanished "good old days." There were winners and losers. There were triumphal accomplishments and abject failures. There were progressive philanthropies for improving the common good and cruel exploitations of the powerless. There were visionary, well-planned achievements and calamitous disasters beyond human control. There was constant, sometimes chaotic, change.

By definition, a work of history looks back. After it fulfills the first and most important principle of recounting history—telling it accurately—the next step is to attempt to answer the question, "what does this mean for us now?"

The manufacturing landscape—and some of the physical landscape—of the Corridor is far different now than it was 100 years ago, or 150 years ago, or 200 years ago. The people here now make their livings differently, live their lives differently, and some even look different than did the Corridor citizens of decades past. The same is true of our nation and world beyond the borders of the five counties of the Corridor. Though future "birthplaces" of yet-unimagined new technologies will surely appear, it is highly unlikely that any place anywhere can ever again be the "world's center" of large-scale manufacture of anything, given the interconnectedness of global business now.

One of the lessons of our history is that major, often traumatic, changes in the economic and industrial life of the Corridor and its residents occurred regularly in the past two centuries. Economic historians such as Alfred Chandler and Walter Licht have written that such disruptions are not only inevitable, but are necessary to bring unsustainable industries or practices to an end. "Some change has been gradual and some cataclysmic, but it's always a disruption," says Don Cunningham, executive director of the Lehigh Valley Economic Development Corporation. "It's always been with us." Whether such disruptions are potentially fatal to an entire place or people, or become a route to a new reality, depends on recognizing that things are always evolving, he says. Having a strong foundation—and recognizing that it exists—are crucial to surviving the disruptions.

"The manufacturing foundation is the most relevant part of our past. It's a mistake to believe that we don't make anything here anymore. In the Lehigh Valley, manufacturing is still one of the top two money generators: we have $7.4 billion in annual output, and about 700 manufacturers employ 32,000 people here. Obviously, it's very different from the Steel era—most manufacturing places employ between 70 and 200 people, but several, such as Mack Trucks, Boston Beer, Nestlé, Alpo, Majestic, B. Braun, and Crayola, have 1,000 or more workers.[1]

"The important factor here is that these places produce a wide variety of hard goods, like packaging materials, medical equipment, electrical components, foods and beverages, cement. We aren't at the mercy of one giant industry that the economy of the whole region relies on," Cunningham pointed out.

The geographical location of the Corridor is still a major factor, he says. "More people have to realize that the shift from relying on natural resources to relying on our excellent access to markets is what makes for success. Forty percent of all U.S. consumers live within an eight-hour truck drive from here." Our mountains and rivers may be more amenities than raw material sources now, but the geographical location of the Corridor's five counties is still pivotal.[*]

The human factor is an equally important foundation, Cunningham believes. "We may not still be making more generations of steel workers or silk weavers, but we still have the work ethic here. There's still a cultural perception and a family culture that accepts and expects that people will be working with their hands and with machinery and making things. And even though many people are now working in our other top industry, which is finance, insurance, and real estate, and our number three industry, education and health care,[2] there is still the work ethic that you do your job to the best of your ability."

The work ethic in the northern counties is no longer quite as baked into the culture, laments Larry Newman, executive director of the Diamond City Partnership in Wilkes-Barrre. "Defeatist, negative attitude is endemic to the anthracite region," he observes. "It's a culture of real loss; so little of what has changed in the last 70 years has been to the good. The negative consequences of the collapse of the anthracite industry have stung longer because so many things that were lost have not been replaced."[3]

But there have long been economic bright spots as well, Newman hastens to add. "When it became apparent that this area was not going to share very much in the economic boom of the 1950s and '60s, communities began to do economic revitalization on their own. They had a radical idea: create industrial parks that could provide attractive and more flexible work spaces for new industries. The parks attracted a host of new industries and now Luzerne County specializes in plastics manufacturing of things like tubing and storage bins. Nardone Brothers is the largest maker of frozen pizza for school food programs, with national distribution. Nabisco, which came to the county when it bought Planters Peanuts, has a large presence here.

"The 'metal bending' industries hark back very clearly to our nineteenth-century industrial heritage," says Newman. "Intermetro Industries makes a huge variety of shelving, carts, and storage units for industry, food service, hospitals and health care." Cornell-Cookson in Mountain Top, which makes steel doors and security grates that are used all over the world, has been a family-owned business since they started a foundry in New York in the 1820s.

---

[*] Among the later major industries to become established in the Corridor were Air Products and Chemicals, Inc., Western Electric (which manufactured the first transistor in its Allentown plant) and Bell Labs, Lutron, OraSure, Just Born, C.F. Martin, Crayola, and Victaulic.

"Even though the industrial park effort was not as successful as was hoped, it worked. The fact is, they are still here and tens of thousands of people have jobs because of the bootstrapping they did in the '50s and '60s."

Like the Lehigh Valley, Luzerne County is seeing rapid growth of logistics, warehousing, and truck transportation. "We are at the upper end of the I-78 and I-81 corridor and the connection to I-84 to New England," Newman points out. "Trucks and interstates are now doing for this area what the railroads did years ago."

Bucks County's industrial areas have been spread out both geographically and in time over the past 200 years. While Durham was one of the first iron furnaces in Pennsylvania in the eighteenth century, and an important iron-industry innovator during the nineteenth century, the county's only other experience with heavy metallurgy was U.S. Steel's Fairless Works from the 1950s to the early '90s. Bristol boasted the largest cluster of manufacturing in the county from the post-Civil War years until the 1960s, producing a wide spectrum of goods from carpet to woven wool to wallpaper to airplanes, and even a space satellite component. Quakertown, Sellersville, and South Langhorne had multiple silk-stocking knitting mills from the early twentieth century up to World War II, when most of them switched to nylon. Silk ribbon was produced in both Doylestown and South Langhorne. There is still a wide range of small manufacturing in Bucks County, ranging from metal fabrication and electronics to musical instruments and food. Langhorne Carpet Company, founded in the early twentieth century, maintains the tradition of weaving fine-quality Wilton carpets. Bucks County is also known for artisan craftwork, such as painting, pottery, weaving, furniture, and jewelry making, which are common throughout the county. Many skilled Bucks County artisans, past and present, are known nationally and even internationally.

Throughout its history, Bucks County has mostly been rural and agricultural. Thus it is unsurprising that two major seed companies, W. Atlee Burpee and Landreth, grew up there. In Solebury Township, the nation's first privately owned soil conservation cooperative, the Honey Hollow Watershed, was established in the 1930s. The five farmers in the area jointly agreed to adopt farming and other techniques such as contour plowing to fight soil erosion and water siltation, practices that are still used and characterize the landscape in the area.

Farmland preservation programs are a major feature of the rural areas of the county today, and residents and visitors alike have a long and often vocal record of opposing many types of development in most areas of the county. It was the combined efforts of local people and part-time, often celebrity, residents that saved and preserved the Delaware Canal.

Biotechnology is a rapidly growing field in all of eastern Pennsylvania, particularly the areas near Philadelphia. Biotech firms are operating in Doylestown, Newtown, New Hope, and Langhorne. Today, the top 50 employers in Bucks County are all in either education or health care, which reflects the proximity and influence of Philadelphia on the region.

The Corridor's third important foundation is an education system that lets people—youthful and mature—develop their abilities. "We have excellent skills education here in the Lehigh

Valley, with three vo-techs and two community colleges that are also very skills-oriented," Don Cunningham points out. "We have more ways than ever before for people to learn new skills to advance in their careers, which we know now is essential for everyone, or to make a mid-career change either out of choice or necessity. We have also found that many of the local people who attend the state colleges come back after graduation, and that's because there are more kinds of jobs here to attract people who have degrees in humanities or creative fields, as well as the ones who have studied finance or the health field."

This reliance on education and skills-building has deep roots. The men described in these pages were mostly not specialists, though some became experts in one field. Most were polymaths, and nearly all self-taught ones at that. Not all had strong formal education, but they were life-long learners. And their belief in education led them to found and support schools of all sorts, from community-based public elementary and high schools to technical schools and universities such as Lehigh, Lafayette, and Wilkes.

Delaware Valley University in Doylestown was founded by an activist Jewish rabbi, Joseph Krauskopf, as an agricultural school to train Jewish immigrants to be farmers, though the school was always open to men of all faiths. Though it has long since become a co-ed liberal-arts university, the college's legacy of linking theoretical training with practical experience is still carried out in a curriculum that requires field work as well as classroom time.

Education and health care are major economic drivers in Wilkes-Barre as well, says Larry Newman. "Half of the 13,000 people who work in downtown Wilkes-Barre are with one of the two universities, Wilkes and King's College, or they work in medicine and health care.

"Both those institutions are the legacies of people who made fortunes in anthracite. They were started to provide higher education to the children of miners. Wilkes was founded in the 1930s as an offshoot of Bucknell. It became independent in 1947, the same year that King's was founded. In recent years, they have each expanded curricula to meet demands for more high-skill career preparation, and as a result we are seeing an upswing in things like tech start-ups, and Wilkes' pharmacy program now has a national reputation. Then there are the innovations that are happening all the time in the Wyoming Valley to clean up acid mine drainage and water pollution, so you could say in a way that anthracite is still driving innovation and job creation."

So geography and human genius are still going strong. What about geology?

According to the Pennsylvania Department of Environmental Protection, 7,637,000 tons of anthracite were produced in 2017, about 5,000 tons less than in 2016. However, only 77,151 tons came out of the 10 remaining underground mines, two in Northumberland County and eight in Schuylkill County. The rest came from either surface mining (3,249,477 tons) or reclamation from coal-waste sites, also known as culm banks (4,310,977 tons). Of the 46 surface mines, three are in Carbon County, and nine in Luzerne; the largest group—22—is found in Schuylkill County. Two coal reclamation firms work in Carbon County, and six in Luzerne. About 900 Pennsylvanians work in the anthracite industry. Anthracite is now mainly used for water purification and sewage treatment, though some is still used in smelting and metal fabrication.

Even the detritus of mining may play a role in the future of U.S. industry. The U.S. Department of Energy (DoE) is funding investigations into the feasibility of recovering rare-earth elements—necessary for high-tech devices such as computer chips and cell phones—from anthracite as well as bituminous waste dumps. The DoE's National Energy Technology Laboratory currently analyzes samples from coalfields in the hope of locating concentrations of the minerals that are high enough to be commercially viable. The federal government considers self-sufficiency in rare-earth elements as critical to national security because virtually all U.S. defense systems are computer-controlled. Since the United States currently imports nearly 100 percent of its rare-earth elements, and most of the world's supply comes from China, developing a domestic supply of these essential materials is a high priority. The future may see an important new industry emerge in the regions where the Corridor's story began more than two hundred years ago.

As for the rivers that were the first routes that transported the fuel of the industrial revolution, they are experiencing a renewal, if not a full rebirth. Since the Clean Water Act of 1972, environmental regulations have limited or ended dumping of untreated sewage, chemical and industrial wastes, and agricultural run-off into rivers, and removed phosphates from laundry detergents. These laws, combined with protections for wetlands and stream banks, have resulted in greatly reduced pollution. Though sand-like coal dirt can still be found on the banks of the Lehigh, the Susquehanna, the Lackawanna, and the Schuylkill, their tributaries and waters no longer run black. Populations of some fish species are resurgent; bald eagles and otters have returned.

Conservation and environmental groups have raised public consciousness about the extent and importance of our watersheds. There is greater awareness that what happens at one point in a waterway has an impact all the way downstream, and even in the ocean. The importance of clean water to the health and economic well-being of people and communities is being more widely recognized and protected.

The rivers of the Corridor—the Lehigh and the Delaware—are no longer transportation routes. Instead, they are now resources for outdoor recreation and tourism and the economic development that these activities bring. Communities on their banks are responding to their residents' desires for more and easier access to the rivers, and cooperating with groups, including the Corridor, that help create trails, bikeways, and boat launches that let people experience the enhanced quality of life that comes from being alongside, on, or in the water.

For centuries, the Lehigh and the Delaware rivers have both shaped the lives of the people living on their banks, and been shaped by them. In 1818, Josiah White and Erskine Hazard paid nothing for the Lehigh River, and created Pennsylvania's economic and industrial empire from it. Now, that priceless treasure and the Delaware River that it feeds are the inheritance of the next generations of the people of the Corridor, preserved for them by benefactors and communities all along the rivers' banks.

Author's Note

No matter whether they are newcomers or members of a family that has lived in this area for decades, people often find it hard to learn the history of their city, town or community. Here are some of my suggestions for getting started.

- Many public libraries have community history collections; more extensive collections can be found in several of the area's college and university libraries.
- The Corridor is fortunate to have many local historical societies, which often have small but interesting collections; some also publish informative newsletters and websites, and invite speakers on local-history and cultural topics.
- Bucks, Northampton, Lehigh, and Luzerne have county historical societies.
- The Heritage Keepers of the Greater Lehigh Valley, formed in 2018, provides information about many historical and preservation organizations through their Facebook page.
- Finally, social media such as the "You know you're from [Bethlehem] if …" Facebook page crowd-source information and bring history-minded people into contact with each other.

MCF

THE CORRIDOR IN COLOR

"View of Mauch-Chunk, Pennsylvania," by Karl Bodmer. This engraving was made in Europe after Bodmer, a Swiss artist, engraver, and trained ethnographer, returned from his travels in the U.S. with German Prince Maximilian of Wied-Neuwied in 1832–1834. Before traveling to the Midwest to study Native American tribes along the Missouri River, Prince Maximilian's companions visited various places in the United States of international renown.

This view of the coal chutes at the foot of the gravity railroad, made from a watercolor painted at the site, is the earliest known contemporary image of Mauch Chunk and coal-loading on the Lehigh.

NCM/D&L

206 ~ Geography, Geology, and Genius

Watercolor by Rufus Grider of the Crane Iron Works, Catasauqua, in 1852. (Moravian Archives)

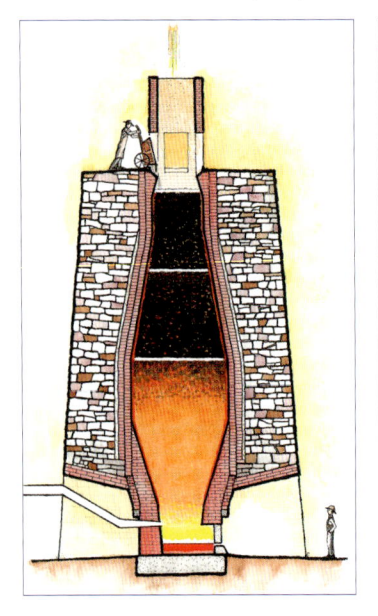

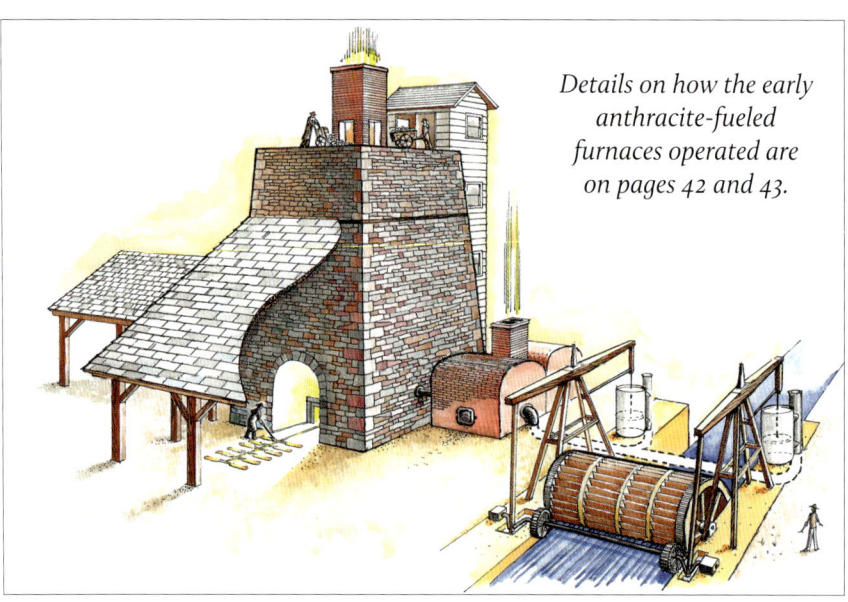

Details on how the early anthracite-fueled furnaces operated are on pages 42 and 43.

Corridor in Color ~ 207

"General View of the Freshet of June 5th 1862 from Mor. Church Steeple taken at 10½ o'clk AM by Rufus A. Grider."

Grider's watercolor shows the flood of 1862 in full spate, with the entire area of the confluence of the Monocacy Creek and the Lehigh River all but submerged. Grider placed small numbers by many of the key features in the painting, and listed them as:

1 – Depot
2 – Canal Bridge
3 – Bridge House
4 – Toll House
5 – Monockacy Bridge
6 – River Bank
7 – Goose Island
8 – Calypso Island
9 – Rice's Boat Yard
10 – West Bethlehem
11 – Borhek & Knauss Store
12 – Keystone Hotel
13 – Water Cure
14 – Fiot, Fountainbleu
15 – Jeter's
16 – Sayre's
17 – L.V.R.R. Offices

All watercolors by Rufus Grider are courtesy of the Moravian Archives in Bethlehem.

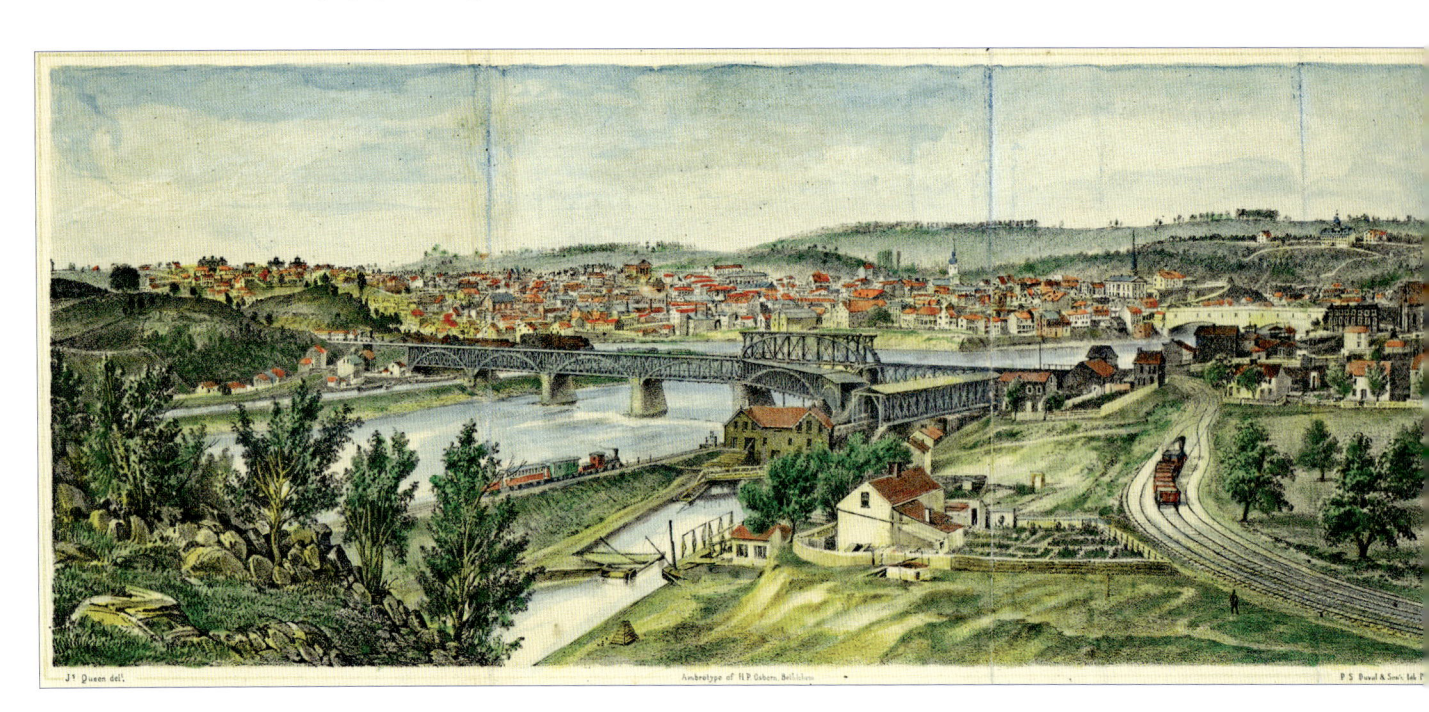

Above is a colorized version of an image that was published in M.S. Henry's 1860 History of the Lehigh Valley. *Easton is in the center, and Phillipsburg on the right. It shows the Forks of the Delaware, where rivers, canals, and railroads came together. Coal came down the Lehigh Navigation on canal boats and continued down the Delaware Division Canal from Easton. Some boats crossed the Delaware River by a tow line to enter the Morris Canal at Phillipsburg, thence to Newark. By this time railroads were competing with the canal system to carry coal, and both the LVRR and the CNJ crossed the Delaware at this point.*

The coal breaker below is typical of innumerable such structures built throughout the coal regions to sort and clean anthracite before it was shipped.

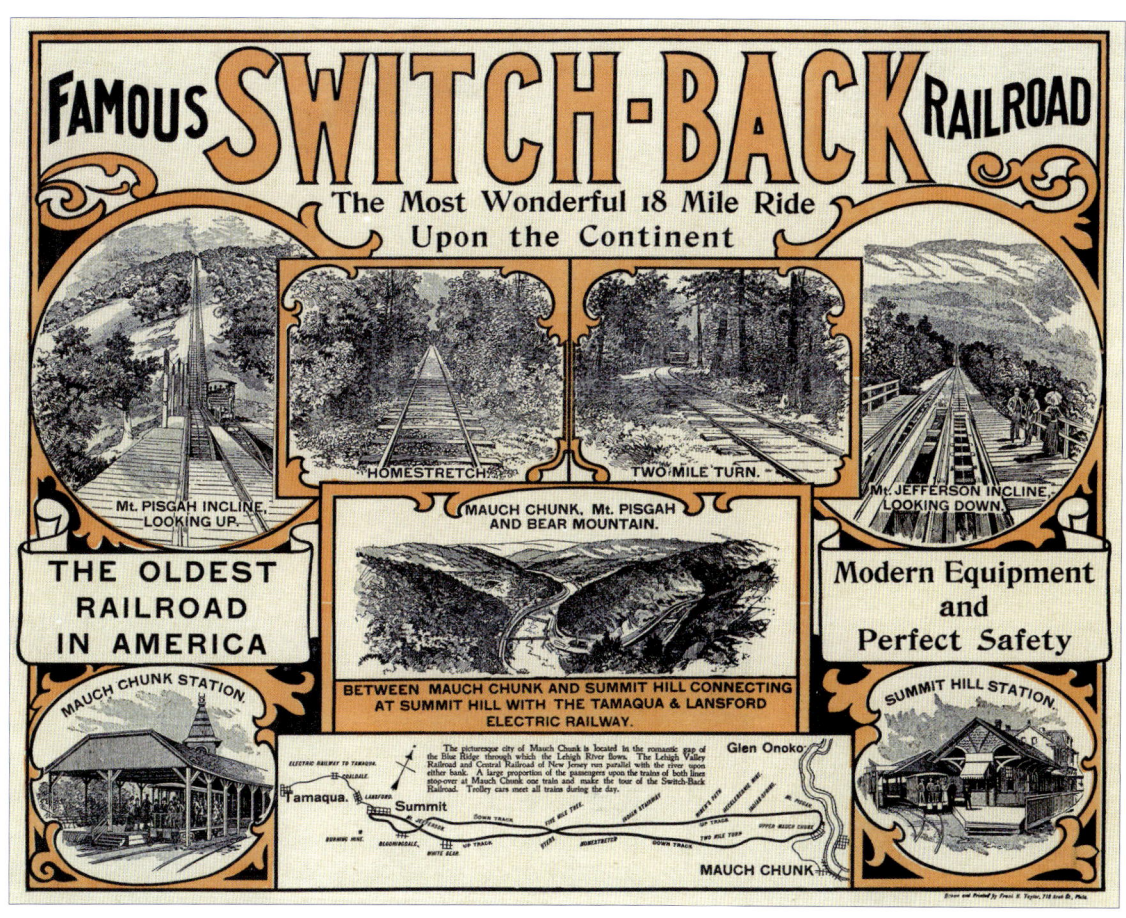

Above, a promotional ad for the Switchback dating from the 1920s. Below, examples of advertising by the Mumford brothers, who ran the Switchback as a tourist attraction 1879–1899.

The logos of many of the cement producers in the Lehigh Valley are displayed here. All the cement was produced from Jacksonburg limestone, which contains the elements needed to produce commercial "portland" cement. The discovery in 1871 of how to produce a manufactured cement for making concrete is one of the most significant and lasting innovations in the Corridor.

(NCM/D&L)

The Adelaide, opened in 1881, was the first of the many silk mills to be built in the Corridor.

(Kate Ruch)

Conrail diesel locomotives No. 6254 and No. 6563, painted in bright blue livery, moving past the blast furnaces at Bethlehem Steel in the 1970s.

(Henry Schmidt for NCM/D&L)

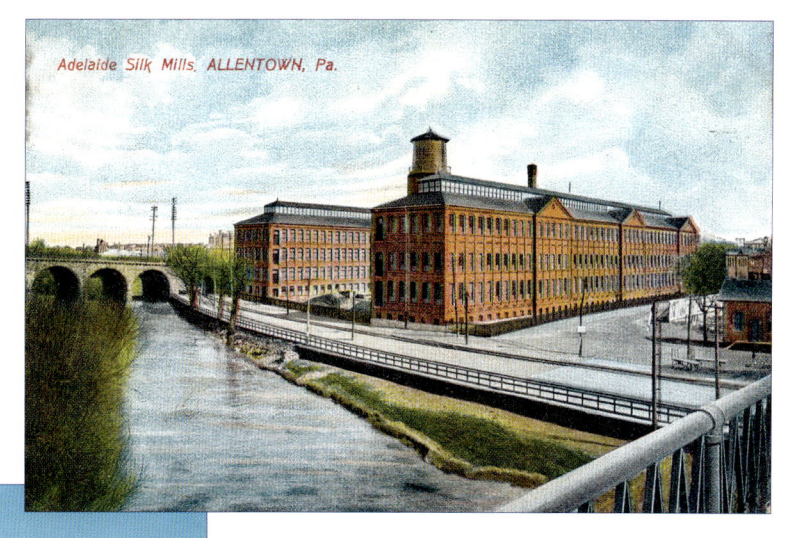

The route of the Lehigh and New England Railroad is marked in red on the map below. This was for many years a vitally important railroad serving the industries of the Corridor from the coal regions to Bethlehem. Its lines were extended to serve primarily the anthracite, cement, and slate industries.

(NCM, Louis J. G. Buehler Collection)

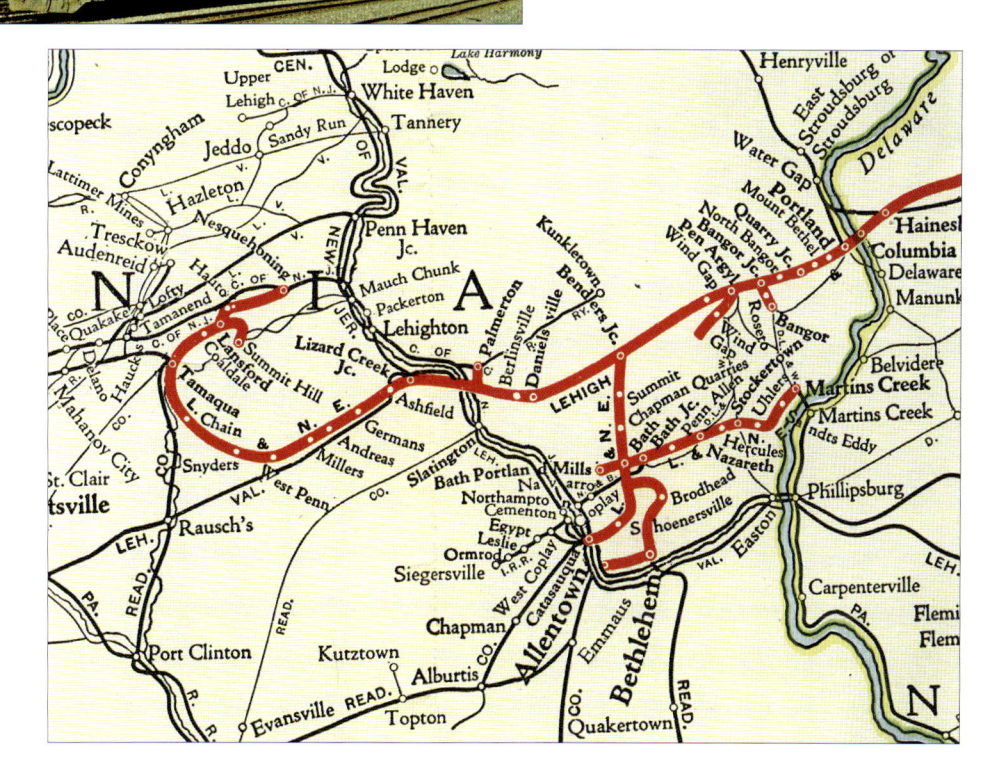

Postcard view of the Individual Drinking Cup's factory in Wilson Brough near Easton, where Dixie cups were made, and a 1930 mockup for a Liberty magazine advertisement for ice cream Dixies.

(Courtesy of Skillman Library, Lafayette College)

The aerial photo below, looking northwest, provides an indication of how large the Bethlehem Steel plant was. It stretched almost 4½ miles along the Lehigh River. This view overlooks the western or Lehigh Division, with the blast furnaces standing tall and prominent near where the company began. There were many more facilities further east along the river.

(NCM/D&L)

Endnotes

The following conventions are used in these notes:

CHTP: Canal History and Technology Press and its predecessor, Center for Canal History and Technology (the publishing department of the National Canal Museum).

HRS: *Historic Resources Study, Delaware and Lehigh Canal National Heritage Corridor and State Heritage Park, 1992.* (This work was published by Hugh Moore Historical Park and Museums, Inc., in 1992 to record the history of the newly created National Heritage Corridor.)

Chapter One, 1790–1819: Coal Ignites the Fire

1. Alfred D. Chandler Jr, *The Visible Hand: The Managerial Revolution in American Business* (Harvard University Press, 1977), 75-78.
2. Ibid., 312.
3. Walter Licht, *Industrializing America: The Nineteenth Century* (Johns Hopkins University Press, 1995), 110-111.
4. H. Benjamin Powell, *Philadelphia's First Fuel Crisis* (Pennsylvania State University Press, 1978), 15.
5. Ibid., 7-28.
6. HRS, 209-210.
7. Josiah White, *Josiah White's History given by himself* (privately printed, Lehigh Coal and Navigation Company, n.d), reprinted 1979 by Carbon County Board of Commissioners, 31.
8. Donald Sayenga, "The Untryed Business: An Appreciation of White and Hazard," *Canal History and Technology Proceedings,* Vol. II, 1983, 105-128.
9. Earl J. Heydinger, "Josiah White and his Bear Trap Navigation," *Canal Currents,* Vol. 41, Winter 1978.
10. Alfred D. Chandler Jr, "Anthracite Coal and the Beginnings of the Industrial Revolution in the United States," *Harvard Business Review,* Vol. XLVI, No. 22, 1972, 150-151.
11. Eleanor Morton, *Josiah White: Prince of Pioneers* (Stephen Daye Press, 1946); *Josiah White's History, given by himself*; Norris Hansell, *Josiah White, Quaker Entrepreneur* (CHTP, 1992).

Chapter Two, 1820–1830: Coal and Canals

1. Alfred D. Chandler Jr, *The Visible Hand: The Managerial Revolution in American Business* (Harvard University Press, 1977), 312.
2. John N. Hoffman, "Anthracite in the Lehigh Region of Pennsylvania, 1820-1845" (Washington, DC: Smithsonian Institution Press, 1968), United States National Museum Bulletin 252, Contributions from The Museum of History and Technology: Paper 72, 93-94.
3. Alfred D. Chandler Jr, "Anthracite Coal and the Beginnings of the Industrial Revolution in the United States," *Harvard Business Review,* Vol. XLVI, No. 22, 1972, 152.
4. Gerald R. Bastoni, "Canvass White, Esquire (1790-1834): Civil Engineer" (Bethlehem, PA: Lehigh University, 1983).

5. Albright G. Zimmerman, *Pennsylvania's Delaware Division Canal: Sixty Miles of Euphoria and Frustration* (Easton, PA: CHTP, 2002), 51.
6. "Annual Report of the Lehigh Coal and Navigation for 1828, 1829," 7; Vincent Hydro Jr, *The Mauch Chunk Switchback: America's Pioneer Railroad* (Easton, PA: CHTP, 2002), 32.
7. H. Benjamin Powell, Philadelphia's First Fuel Crisis, 97.

Chapter Three, 1830–1840: Connecting Mine to Market

1. Donald Sayenga, "The Untryed Business: An Appreciation of White and Hazard," *Canal History and Technology Proceedings,* Vol. II, 1983, 115.
2. Sayenga, "Untryed Business, 114–117.
3. Albright G. Zimmerman, *Pennsylvania's Delaware Division Canal: Sixty Miles of Euphoria and Frustration* (Easton, PA: CHTP, 2002), 31.
4. Robert Kapsch, *Over the Alleghenies* (University of Pittsburgh Press, 2013), 287–290.
5. Zimmerman, *Delaware Division Canal*, 32–33.
6. HRS, 145.
7. Quoted in "King of Coal Barons, Father of Hazleton," *The Standard-Speaker*, 16 Jan 2016.
8. HRS, 161–163, 169–170.
9. F. Charles Petrillo, *Anthracite and Slackwater: The North Branch Canal, 1828–1901* (Easton, PA: CHTP, 1986), 53.
10. HRS, 213.
11. Ibid., 211–212.
12. Alfred D. Chandler Jr, "Anthracite Coal and the Beginnings of the Industrial Revolution in the United States," *Harvard Business Review*, Vol. XLVI, No. 22, 1972, 156–158, 179–180.

Chapter Four, 1840–1860: Revolutionary Anthracite Iron

1. Alfred D. Chandler Jr, The Visible Hand: The Managerial Revolution in American Business (Harvard University Press, 1977), 312.
2. Craig L. Bartholomew and Lance E. Metz, *The Anthracite Iron Industry of the Lehigh Valley* (Easton, PA: CHTP, 1988), 11–12.
3. HRS, 226.
4. Bartholomew and Metz, *Anthracite Iron Industry*, 19.
5. Ibid., 15–19.
6. Ibid., 22–23.
7. HRS, 228.
8. Bartholomew and Metz, *Anthracite Iron Industry*, 20–24.
9. Ibid., 25, 27.
10. Ibid., 27.
11. Ibid.
12. Ibid.
13. Ibid.
14. William Firmstone, "Sketch of Early Anthracite Furnaces," *Transactions of the American Institute of Mining Engineers*, Vol. III, 1874, 155–156.
15. *Catasauqua Dispatch*, undated clipping in Catasauqua Library.

Chapter 5: 1840–1850: The Iron Age

1. Craig L. Bartholomew and Lance E. Metz, *The Anthracite Iron Industry of the Lehigh Valley* (Easton, PA: CHTP, 1988), 52.
2. HRS, 231–232.
3. Bartholomew and Metz, *Anthracite Iron Industry*, 130–131.
4. HRS, 248–249.
5. James F. Lambert and Henry Reinhard, *A History of Catasauqua in Lehigh County, Pennsylvania* (Allentown, PA: The Searle and Dressler Co., 1914), 348.
6. HRS, 237–240.
7. Ibid., 213–214
8. M.S. Henry, *History of the Lehigh Valley* (Easton, PA: Bixler & Corwin, 1860), 366–367; www.gendisasters.com/pennsylvania/6320/stockton-pa-collapse-coal-mine-dec-1869
9. HRS, 213–215.
10. Charles E. Peterson, "The Spider Bridge, A Curious Work at the Falls of Schuylkill, 1816," *Canal History and Technology Proceedings*, Vol. V, 1986. 243–259.
11. Donald Sayenga, "The Early Years of America's Wire Rope Industry 1818–1848," *Canal History and Technology Proceedings*, Vol. X, 1991, 149–80; www.bridonamerican.com/us.

Chapter 6: 1850–1860: Mineral Wealth

1. Alfred D. Chandler Jr, *The Visible Hand: The Managerial Revolution in American Business* (Harvard University Press, 1977), 76–77.
2. HRS, 214.
3. Lynn Brubaker, "The Wapwallopen Mills 1859–1865" (unpublished 1961 research paper, on deposit at Hagley Museum and Archives, Wilmington, DE, and National Canal Museum Archives, Easton, PA), 2.
4. Ibid., 4.
5. HRS, 216.
6. Ibid., 216–217.
7. Robert P. Wolensky and Joseph M. Keating, *Tragedy at Avondale* (Easton, PA: CHTP, 2008) offers a comprehensive discussion of the nature of the disaster and the trial and the conflicting evidence, with many contemporary illustrations.
8. Ibid., 99–104.
9. Ibid., 105–108.
10. https://en.wikipedia.org/wiki/History_of_immigration_to_the_United_States.
11. Donald L. Miller and Richard E. Sharpless, *The Kingdom of Coal: Work, Enterprise, and Ethnic Communities in the Mine Fields* (Philadelphia: University of Pennsylvania Press, 1985; reprint, CHTP, 1998), 140–142.
12. James F. Lambert and Henry J. Reinhard, *A History of Catasauqua in Lehigh County Pennsylvania* (Allentown, PA: The Searle and Dressler Co., 1914), 10.
13. O.P. Knauss, *Millerstown* (Macungie, PA: O.P. Knauss, 1943), 18–19.
14. Alfred Mathews and Austin N. Hungerford, *History of the Counties of Lehigh and Carbon in the Commonwealth of Pennsylvania* (Philadelphia: Everts & Richards, 1884), 328–329.
15. Albert Ohl, *History of Southern Lehigh County, Pa., 1732–1947* (Coopersburg, PA: Lions Club, bound volume of Ohl's handwritten account, 1952), 112–113.
16. HRS, 242.
17. Craig L. Bartholomew and Lance E. Metz, *The Anthracite Iron Industry of the Lehigh Valley* (Easton, PA: CHTP, 1988), 110–111.

18. Ibid., 145–47.
19. Dr. John Percy, *Metallurgy: Iron and Steel* (London, 1864) wrote this in his chapter on furnaces in America. The Thomas Iron Company included the quote in *The Thomas Iron Company, 1854–1904* (self-published, 1904), 67.
20. J.P. Lesley, *The Iron Manufacturer's Guide to the Furnaces, Forges and Rolling Mills of the United States* (New York: John Wiley, 1866), 9.
21. Bartholomew and Metz, *Anthracite Iron Industry*, 167–169.
22. HRS, 313–314.

Chapter 7, 1860–1870: Iron Horses and Iron Rails

1. HRS, 258–259.
2. Lance Metz, *Robert H. Sayre: Engineer, Entrepreneur, Humanist, 1824–1907* (Easton, PA: Hugh Moore Historical Park, 1985), pamphlet.
3. HRS, 258–259.
4. Lance E. Metz and Donald Sayenga, "The Role of John Fritz in the Development of the Three-High Rail Mill, 1855–1863," paper published for the Ironmasters Conference, Lehigh Valley, PA, 2001, 41–62.
5. Lance E. Metz, *John Fritz 1822–1913: His Role in the Development of the American Iron and Steel Industry* (Easton, PA: CHTP, 1988), pamphlet.
6. HRS, 259–260; Lance Metz, *John Fritz: His Role in the Development of the American Iron and Steel Industry and His Legacy to the Bethlehem Community* (Hugh Moore Historical Park and Museums, Inc., and Annie S. Kemerer Museum, 1987), 22.
7. Joan Gilbert, *Gateway to the Coal Fields: The Upper Grand Section of the Lehigh Canal* (Easton, PA: CHTP, 2005), 137–138.
8. Gilbert, *Gateway,* 137–140; *Report of the Board of Managers of the Lehigh Coal and Navigation Company to the Stockholders, May 5, 1863,* 42–43.
9. Joseph Levering, *A History of Bethlehem, Pennsylvania, 1741–1892* (Bethlehem, PA: Times Publishing Company, 1903), 736.
10. http://www.weather.gov/media/marfc/Flood_Events_2016/1862/Jun6%2C1862.pdf
11. *Incidents of the Freshet on the Lehigh River, Sixth Month, 4th and 5th, 1862* (Crissy & Markley, Printers, 1863), 24–25.
12. www.findagrave.com/cgi-bin/fg.cgi?page=gr&GRid=28920486
13. Alfred Mathews and Austin N. Hungerford, *History of the Counties of Lehigh and Carbon in the Commonwealth of Pennsylvania* (Philadelphia: Everts & Richards, 1884; reproduced by Lehigh Gap Historical Society, Palmerton, 1995), 685.
14. HRS, 161–167.

Chapter 8, 1870–1880: Rise and Fall and Rise

1. HRS, 243–245.
2. Craig L. Bartholomew and Lance E. Metz, *The Anthracite Iron Industry of the Lehigh Valley* (Easton, PA: CHTP, 1988), 112–113.
3. Ibid., 124–126.
4. Ibid., 145–147.
5. Ibid., 163–164.
6. Lance E. Metz, *John Fritz 1822–1913* (Easton, PA: CHTP, 1988); John Fritz, *The Autobiography of John Fritz* (John Wiley & Sons, 1912).
7. Robert W. Hunt, "A History of Bessemer Manufacture in America," *The Transactions of the American Institute of Mining Engineers*, Vol. 5, (1876–1877), 212–213.

8. HRS, 261.
9. Ibid., 260-262.
10. Ibid., 213-215.
11. Benjamin F. Fackenthal, "The Manufacture of Hydraulic Cements in Bucks County," *Papers Read before the Bucks County Historical Society,* vol. 6, 1930.
12. Martha Capwell Fox, "From Sugar to Cement: The Long and Varied Entrepreneurial Career of José de Navarro," *Canal History and Technology Proceedings,* Vol. XXIX, 2010, 152-154.
13. HRS, 306.
14. Nadine Miller Peterson and Dan Zagorski, "Zinc Mining in the Saucon Valley Region of Pennsylvania 1846-1986," *Canal History and Technology Proceedings,* Vol. XX, 2001, 139-162.
15. Ibid.
16. Historical Collection, Margaret M. Grundy Library, Bristol, PA.

Chapter 9, The 1880-1890: Steel, Cement—and Silk

1. HRS, 263-268.
2. Robert Lesley, *History of the Portland Cement Industry in the United States* (International Trade Press, 1924), 60-66.
3. Martha Capwell Fox, "From Sugar To Cement: The Long and Varied Entrepreneurial Career of José de Navarro," *Canal History and Technology Proceedings*, Vol. XXIX, 2010, 139-157.
4. Lesley, *Portland Cement Industry*, 110.
5. Fox, "From Sugar To Cement," 139-157.
6. Ken Bloom and Marion Wolbers, *Allentown, A Pictorial History* (Virginia Beach: Donning Co., 1984), 51.
7. Martha Capwell Fox, "Seams of Coal, Beams of Steel, Skeins of Silk: The Silk Industry in the Delaware and Lehigh National Heritage Corridor," *Canal History and Technology Proceedings*, Vol. XXI, 2002, 85-96.
8. John F. Sears, *Sacred Places: American Tourist Attractions in the Nineteenth Century* (New York: Oxford University Press, 1989; repr., University of Massachusetts Press, 1999), 191.
9. Much of this section has been taken with permission from Vince Hydro, Jr., *The Mauch Chunk Switchback: America's Pioneer Railroad* (Easton, PA: CHTP, 2002), chapters 6, 9.

Chapter 10, 1890-1900: Armor, Immigration, and Cement

1. HRS, 262-278.
2. Ann Bartholomew and Donald Stuart Young, *Bethlehem Steel in Bethlehem, Pennsylvania: a photographic history* (Easton, PA: CHTP, 2010), 23.
3. *Allentown Daily Leader*, 21 Oct 1899, 1.
4. *Allentown Daily Leader* 18 Jan 1896, 2.
5. Frank A. Whelan, "'I Can Stand It': Charles M. Schwab and the Bethlehem Steel strike of 1910," *Canal History and Technology Proceedings*, Vol. XIII, 1994, Appendix, 77-80.
6. Richard D. Grifo and Anthony F. Noto, *Italian Presence in Pennsylvania* (University Park, PA: Pennsylvania Historical Association, 1990).
7. Robert P. Wolensky and William A. Hastie Sr., *Anthracite Labor Wars* (Easton, PA: CHTP, 2013), 216, 234.
8. HRS, 217-218.
9. Joan Minton Christopher, Martha Capwell Fox, and Carol Front, *Cement Industry of the Lehigh Valley* (Charleston, SC: Arcadia Publishing, 2006).
10. *A Story Runs Through It: Wyoming Valley Levee System* (Luzerne County Flood Protection Agency in collaboration with the Luzerne County Historical Society, 2003), 88.

11. James J. Bohning, "Angel of the Anthracite: The Philanthropic Legacy of Sophia Georgina Coxe," *Canal History and Technology Proceedings*, Vol. XXIV, 2005, 150-182.
12. HRS, 315.
13. Ibid.

Chapter 11, 1900-1910: New Century, New Industries, and a New Name

1. HRS, 316; Joan Campion, *Smokestacks and Black Diamonds: A History of Carbon County, Pennsylvania* (CHTP, 1997), 201-214.
2. John B. Montville, *Mack* (Newfoundland, N.J: Haessner Publishing, 1973), 22-23.
3. Ibid., 17.
4. Ibid., 25.
5. HRS, 278-279.
6. Joan Campion, *Saturday Night on the South Side* (South Bethlehem Historical Society, 1988), 1, 6.
7. Ibid.
8. HRS, 279; Robert Hessen, *Steel Titan* (New York: Oxford University Press, 1975), 167.
9. HRS, 278-281.
10. Anne Hawkes Hutton, *The Pennsylvanian—Joseph Grundy* (Pittsburgh, PA: Dorrance and Co., 1982) Hutton, *Grundy*; www.explorepahistory.com/hmarker.php?markerId=1-A-364.
11. Hutton, *Grundy*; Historical Collection, Margaret M. Grundy Library, Bristol, PA.
12. Robert Janosov, "Beyond 'The Great Coal Strike': The Anthracite Coal Region in 1902," paper published in *The "Great Strike": Perspectives on the 1902 Anthracite Coal Strike* (Easton, PA: CHTP, 2002), 1-28.
13. George O. Virtue, "The Anthracite Mine Laborers," *Bulletin of the Department of Labor*, Vol. 2, No. 13, Nov 1897.
14. Donald L. Miller and Richard E. Sharpless, The Kingdom of Coal: Work, Enterprise, and Ethnic Communities in the Mine Fields (Philadelphia: University of Pennsylvania Press, 1985; reprint, Canal History and Technology Press, 1998), 260.
15. From Ohio State University's compilation of political cartoons about the coal strike. This one can be found at https://ehistory.osu.edu/exhibitions/gildedage/1902AnthraciteStrike/content/Baer
16. Miller and Sharpless, *Kingdom of Coal*, 277-279.
17. Ibid., 280-282.

Chapter 12, 1910-1920: Labor War and World War

1. Frank Whelan, "'I can stand it': Charles M. Schwab and the Bethlehem Steel Strike of 1910" *Canal History and Technology Proceedings*, Vol. XIII, 1994, 63-80.
2. Kenneth Warren, *Industrial Genius: The Working Life of Charles Michael Schwab* (University of Pittsburgh Press, 2007), 125.
3. Ann Bartholomew, "Allentown, 1917-1920," in Mahlon H. Hellerich, ed., *Allentown 1762-1987: A 225-year History* (Allentown, PA: Lehigh County Historical Society, 1987), Vol. 1, 447-453; Louis Rodriquez, *From Elephants to Swimming Pools: Carl Akeley, Samuel W. Traylor, and the Development of the Cement Gun* (Easton, PA: CHTP, 2006), 53-54.
4. Bartholomew, "Allentown, 1917-1920," Vol. 1, 500; Ann Bartholomew and Carol M. Front, *Allentown* (Arcadia Publishing, Images of America series, 2002) 14; Frank Whelan, "Wire Mill Once Thrived," *The Morning Call*, 13 April 2005.
5. https://archive.org/stream/historyofmanufac00will/historyofmanufac00will_djvu.txt
6. "Adelaide Mill Gets Government Contract," *The Allentown Morning Call*, 21 July 1919, 5; Bartholomew, "Allentown 1917-1920," Vol. 1, 540.
7. John B. Montville, *Mack* (Newfoundland, N.J: Haessner Publishing, 1973), 27-30, 34-36.

8. Ibid., 56.
9. Ibid., 96.
10. Montville, *Mack*, 51-58.
11. Rodriquez, *Elephants to Swimming Pools*, 55-56.
12. www.globalsecurity.org/military/facility/bristol.htm and www.livingplaces.com/PA/Bucks_County/Bristol_Borough/Harriman.html

Chapter 13, 1920–1930: The Roaring Twenties

1. Michael Taylor, *Jane's Encyclopedia of Aviation,* Studio Editions, 560; "Another Transatlantic Plane Crash, 1927," www.britishpathe.com/video/another-transatlantic-plane-crash/query/01071000.
2. Curtiss-Wright Sales Corporation, *Keystone Aircraft Corporation, Division of Curtiss-Wright, undated.*
3. Anke Voss-Hubbard, "Hugh Moore and the Selling of the paper Cup: A History of the Dixie Cup Company 1907–1957," *Canal History and Technology Proceedings*, Vol. XV, 1996, 87.
4. Ibid., 95.
5. HRS, 258.
6. Craig L. Bartholomew and Lance E. Metz, *The Anthracite Iron Industry of the Lehigh Valley* (Easton, PA: CHTP, 1988), 121–122.
7. Kenneth Warren, *Industrial Genius: The Working Life of Charles Michael Schwab* (Pittsburgh, PA: University of Pittsburgh Press, 2007), 122.
8. Ann Bartholomew, "Allentown: 1917–1920" in Mahlon H. Hellerich, ed., *Allentown 1762–1987: A 225-Year History* (Allentown, Lehigh County Historical Society, 1987), vol. 1, 511–512.
9. Martha Capwell Fox, "Seams of Coal, Beams of Steel, Skeins of Silk: The Silk Industry in the Delaware and Lehigh National Heritage Corridor," *Canal History and Technology Proceedings*, Vol. XXI, 2002, 85–96.
10. Fox, "Silk Industry."
11. Roberta Burkhardt and Judith Gemmel, *Catasauqua and North Catasauqua: A Profile of the Boroughs* (Historic Catasauqua Preservation Association, 1992), 145–146.
12. There does not appear to be a definitive answer to the question of how many miles of paved roads there were in the U.S. in 1920.
13. Joseph S. Young, *A Brief Outline of the History of Cement* (Allentown, PA: Lehigh Portland Cement, 1942), 20–24.
14. Joan Minton Christopher, Martha Capwell Fox, and Carol Front, *Cement Industry of the Lehigh Valley* (Charleston, SC: Arcadia Publishing, 2006), 28, 44–45.
15. Louis Rodriquez, *From Elephants to Swimming Pools: Carl Akeley, Samuel W. Traylor, and the Development of the Cement Gun* (Easton, PA: CHTP, 2006), 41.
16. Burkhardt and Gemmel, 29–33.
17. John Heilig, *Mack: Driven for a Century* (Motorbooks International, 1999), 30.
18. John B. Montville, *Mack* (Newfoundland, N.J: Haessner Publishing, 1973), 66–68, 82–83.
19. HRS, 218.
20. Ben Marsh, "Continuity and Decline in the Anthracite Towns of Pennsylvania," *Annals of the Association of American Geographers,* Vol. 77, no. 3, 337–352; Thomas Dublin and Walter Licht, *The Face of Decline: The Pennsylvania Anthracite Region in the Twentieth Century* (Ithaca, NY: Cornell University Press), 2005.
21. Pennsylvania Population (2018-01-19). Retrieved 2018-07-26, from http://worldpopulationreview.com/states/pennsylvania-population/
22. https://www.eia.gov/coal/annual/pdf/table18.pdf

Chapter 14: 1930–1960: Depression, War, Booms, and Busts

1. John B. Lovis, *Steelmaking in Bethlehem, PA: The Final Years* (Easton, PA: CHTP, 2015), 2, 16–17, 19.
2. www.lhforge.com.
3. Joseph S. Young, "Fifty Years of Progress" (Allentown, PA: Lehigh Portland Cement Company, 1947), 17.
4. www.phmc.state.pa.us/portal/communities/pa-heritage/files/disaster-murder-mines.pdf.
5. Pennsylvania Department of Environmental Protection, Bureau of Mining Programs, 2016.
6. John B. Montville, *Mack* (Newfoundland, N.J: Haessner Publishing, 1973), 119–123; John Heilig, *Mack: Driven for a Century* (Motorbooks International, 1999), 72–84.
7. https://www.macktrucks.com/trucks/anthem/
8. Miscellaneous company publications and newspaper articles, Grundy Library, Bristol, PA.
9. www.conrail.com/history; William C. Vantuono, "Conrail at 40: An Experiment that Worked," *Railway Age*, 5 April 2016; www.railwayage.com/freight/class-i/conrail-at-40-the-experiment-still-works/
10. www.nscorp.com/content/dam/nscorp/get-to-know-ns/investor-relations/annual-reports/annual-report-2017.pdf
11. Albright G. Zimmerman, *Pennsylvania's Delaware Division Canal: Sixty Miles of Euphoria and Frustration* (Easton, PA: CHTP, 2002), 126.

Epilogue

1. Conversation, author with Don Cunningham, 23 Jan 2018.
2. Ibid. Cunningham's regular column in the Sunday business section of *The Morning Call* provides a great deal of current economic data.
3. Conversation, author with Larry Newman, 23 April 2018.

Bibliography

The following convention is used in this bibliography: CHTP for Canal History and Technology Press and its original name, Center for Canal History and Technology.

Allentown Morning Call. "Adelaide Mill Gets Government Contract." 21 July 1919.

Allentown Daily Leader. 21 Oct 1899; 18 Jan 1896.

Annual Report of Lehigh Coal and Navigation for 1828, 1829.

Bartholomew, Ann. "Allentown 1917-1920." In *Allentown 1762-1987: A 225-year History,* edited by Mahlon H. Hellerich. Allentown, PA: Lehigh County Historical Society, 1987, Vol. 1.

Bartholomew, Ann, and Carol M. Front. *Allentown.* Charleston, SC: Arcadia Publishing, *Images of America* series, 2002.

Bartholomew, Ann, and Donald Stuart Young. *Bethlehem Steel in Bethlehem, Pennsylvania: a photographic history.* Easton, PA: CHTP, 2010.

Bartholomew, Ann, and Lance E. Metz. *Delaware and Lehigh Canals: A Pictorial History of the Delaware and Lehigh Canals National Heritage Corridor in Pennsylvania.* Easton, PA: CHTP, second edition, 2005.

Bartholomew, Craig L., and Lance E. Metz. *The Anthracite Iron Industry of the Lehigh Valley.* Easton, PA: CHTP, 1988.

Bastoni, Gerald R. "Canvass White, Esquire (1790-1834): Civil Engineer." Bethlehem, PA: Lehigh University, 1983.

Blalock, Thomas J. "The Bethlehem Steel Works During the Early Twentieth Century and the Development of its Electric Power System." *Canal History and Technology Proceedings* Vol. XXIV, 2005.

Bloom, Ken, and Marion Wolbers. *Allentown, A Pictorial History.* Virginia Beach: Donning Co., 1984.

Bohning, James J. "Angel of the Anthracite: The Philanthropic Legacy of Sophia Georgina Coxe." *Canal History and Technology Proceedings* Vol. XXIV, 2005.

Bowen, Eli. *The Pictorial Sketch-Book of Pennsylvania.* Philadelphia: Willis P. Hazard, 1852.

Brubaker, Lynn. "The Wapwallopen Mills 1859-1865." Unpublished 1961 research paper. On deposit at Hagley Museum and Archives, Wilmington, DE, and National Canal Museum Archives, Easton, PA.

Bryant, William Cullen, *Picturesque America.* D. Appleton and Co., New York, 1872.

Burkhardt, Roberta, and Judith Gemmel. *Catasauqua and North Catasauqua: A Profile of the Boroughs.* Historic Catasauqua Preservation Association, 1992.

Campion, Joan. *Saturday Night on the South Side.* South Bethlehem Historical Society, 1988.

———. *Smokestacks and Black Diamonds: A History of Carbon County, Pennsylvania.* CHTP, 1997.

Chandler, Alfred D. Jr. "Anthracite Coal and the Beginnings of the Industrial Revolution in the United States." *Harvard Business Review* Vol. XLVI, No. 22, 1972.

———. *The Visible Hand: The Managerial Revolution in American Business.* Harvard University Press, 1977.

Christopher, Joan Minton, Martha Capwell Fox, and Carol Front. *Cement Industry of the Lehigh Valley.* Charleston, SC: Arcadia Publishing, *Images of America* series 2006.

DeNie, Wouter. "Lehigh Canal." Unpublished draft study in the collections of Hugh Moore Park. 1998.

Dublin, Thomas, and Walter Licht. *The Face of Decline: The Pennsylvania Anthracite Region in the Twentieth Century.* Ithaca, NY: Cornell University Press, 2005.

Fackenthal, Benjamin F. "The Manufacture of Hydraulic Cements in Bucks County" *Bucks County Historical Society,* vol. 6, 1930.

Firmstone, William. "Sketch of Early Anthracite Furnaces." *Transactions of the American Institute of Mining Engineers.* Vol, III, 1874.

Fox, Martha Capwell. "From Sugar to Cement: The Long and Varied Entrepreneurial Career of José de Navarro." *Canal History and Technology Proceedings,* Vol. XXIX, 2010.

———. "Seams of Coal, Beams of Steel, Skeins of Silk: The Silk Industry in the Delaware and Lehigh National Heritage Corridor," *Canal History and Technology Proceedings*, Vol. XXI, 2002.

Gilbert, Joan. *Gateway to the Coal Fields: The Upper Grand Section of the Lehigh Canal.* Easton, PA: CHTP, 2005.

Grifo, Richard D., and Anthony F. Noto. *Italian Presence in Pennsylvania*. University Park, PA: Pennsylvania Historical Association, 1990.

Grundy, Joseph. "Industrial Growth of Bristol Borough." Papers Read Before the Bucks County Historical Society, Vol. IV, 1917.

Hansell, Norris. *Josiah White, Quaker Entrepreneur.* CHTP, 1992.

Heilig, John. *Mack: Driven for a Century.* Motorbooks International, 1999.

Henry, M. S. *History of the Lehigh Valley.* Easton, PA: Bixler and Corwin, 1860.

Hessen, Robert. *Steel Titan.* New York: Oxford University Press, 1975.

Heydinger, Earl J. "Josiah White and his Bear Trap Navigation" *Canal Currents*, Vol. 41, Winter 1978.

Historic Resources Study, Delaware and Lehigh Canal National Heritage Corridor and State Heritage Park, 1992. Hugh Moore Historical Park and Museums [CHTP]. 1992.

Historical Collection, Margaret M. Grundy Library, Bristol, PA.

Hoffman, John N. "Anthracite in the Lehigh Region of Pennsylvania, 1820-1845." Smithsonian Institution Press, 1968.

Holleran, Philip M. "Explaining the decline of child labor in Pennsylvania silk mills, 1899-1919." *Pennsylvania History*, Vol. 63, No. 1, Winter 1996.

Hunt, Robert W. "A History of Bessemer Manufacture in America." *Transactions of the American Institute of Mining Engineers*, Vol. 5, 1876-1877.

Hutton, Anne Hawkes. *The Pennsylvanian—Joseph Grundy.* Pittsburgh, PA: Dorrance and Co., 1982.

Hydro, Vincent Jr. *The Mauch Chunk Switchback: America's Pioneer Railroad*. CHTP, 2002.

Incidents of the Freshet on the Lehigh River, Sixth Month, 4th and 5th, 1862. Crissy & Markley, Printers, 1863.

Janosov, Robert. "Beyond 'The Great Coal Strike': The Anthracite Coal Region in 1902." *The "Great Strike": Perspectives on the 1902 Anthracite Coal Strike."* Easton, PA: CHTP, 2002.

Kapsch, Robert. *Over the Alleghenies.* University of Pittsburgh Press, 2013.

Knauss, O.P. *Millerstown.* Macungie, PA: O.P. Knauss, 1943.

Knies, Michael. *Coal on the Lehigh1790-1827: Beginnings and Growth of the Anthracite Industry in Carbon County, Pennsylvania.* Easton, PA: CHTP, 2001.

Kulp, Randolph L., editor. *History of the Lehigh and New England Railroad Company.* Allentown, PA: Lehigh Valley Chapter, National Railway Historical Society, Inc., 1972.

Lambert, James F., and Henry Reinhard. *A History of Catasauqua in Lehigh County, Pennsylvania.* Catasauqua, PA, 1914.

Lesley, Robert. *History of the Portland Cement Industry in the United States.* International Trade Press, 1924.

Lesley, J.P. *The Iron Manufacturer's Guide to the Furnaces, Forges and Rolling Mills of the United States.* New York: John Wiley, 1866.

Levering, Joseph. *A History of Bethlehem, Pennsylvania, 1741-1892.* Bethlehem, PA: Times Publishing Company, 1903.

Licht, Walter. *Industrializing America: The Nineteenth Century.* Johns Hopkins University Press, 1995.

Lovis, John B. *Steelmaking in Bethlehem, PA: The Final Years.* Easton, PA: CHTP, 2015.

Lower Macungie Bicentennial Committee. *A History of Lower Macungie Township, Lehigh County, Pennsylvania.* Lower Macungie Township, PA: Lower Macungie Bicentennial Committee, 1976; second edition, Lower Macungie Township Historical Society, 1996.

Luzerne County Flood Protection Agency in collaboration with the Luzerne County Historical Society. *A Story Runs Through It: Wyoming Valley Levee System.* 2003.

Marsh, Ben. "Continuity and Decline in the Anthracite Towns of Pennsylvania." *Annals of the Association of American Geographers* Vol. 77, no. 3.

Mathews, Alfred, and Austin N. Hungerford. *History of the Counties of Lehigh and Carbon in the Commonwealth of Pennsylvania.* Philadelphia: Everts & Richards, 1884.

Metz, Lance. *John Fritz 1822-1913: His Role in the Development of the American Iron and Steel Industry.* Easton, PA: CHTP, 1988.

———. *Robert H. Sayre: Engineer, Entrepreneur, Humanist, 1824-1907.* Easton, PA: CHTP, 1985.

Metz, Lance, and Donald Sayenga. "The Role of John Fritz in the Development of the Three-High Rail Mill, 1855-1863." Paper presented at the Ironmasters Conference, Lehigh Valley, PA, 2001.

Miller, Benjamin L. *Lehigh County, Pennsylvania, Geology and Geography.* Harrisburg, PA: Dept. of Internal Affairs, 1941.

———, Donald M. Fraser, and Ralph L. Miller. *Pennsylvana Geological Survey: Northampton County.* Harrisburg, PA: Commonwealth of Pennsylvania, 1939.

Miller, Donald L., and Richard E. Sharpless. *The Kingdom of Coal: Work, Enterprise, and Ethnic Communities in the Mine Fields.* Philadelphia: University of Pennsylvania Press, 1985; reprint, CHTP, 1998.

Mitchener, Harold, and Carol Mitchener. *Bristol.* Charleston, SC: Arcadia Publishing, *Images of America* series, 2000.

Montville, John B. *Mack.* Newfoundland, N.J: Haessner Publishing, 1973.

Morton, Eleanor. *Josiah White: Prince of Pioneers.* Stephen Daye Press, 1946.

Munsell, W.W. & Co. *History of Luzerne, Lackawanna and Wyoming Counties, Pa.* New York, 1880.

Ohl, Albert. *History of Southern Lehigh County, Pa., 1732-1947.* Coopersburg, PA: Lions Club. Bound volume of Ohl's handwritten account, 1952.

Pennsylvania Department of Environmental Protection, Bureau of Mining Programs, 2016.

Peterson, Charles E. "The Spider Bridge, A Curious Work at the Falls of Schuylkill, 1816." *Canal History and Technology Proceedings.* Vol. V, 1986.

Peterson, Nadine Miller, and Dan Zagorski, "Zinc Mining in the Saucon Valley Region of Pennsylvania 1846-1986." *Canal History and Technology Proceedings* Vol. XX, 2001.

Petrillo, F. Charles. *Anthracite and Slackwater: The North Branch Canal, 1828-1901.* Easton, PA: CHTP, 1986.

Powell, H. Benjamin. *Philadelphia's First Fuel Crisis.* Pennsylvania State University Press, 1978.

Richardson, Richard. *A Memoir of Josiah White.* Philadelphia: J. B. Lippincott and Co., 1873.

Rodriquez, Louis. *From Elephants to Swimming Pools: Carl Akeley, Samuel W. Traylor, and the Development of the Cement Gun.* Easton, PA: CHTP, 2006.

Rupp, I. Daniel. *History of Northampton, Lehigh, Monroe, Carbon, and Schuylkill Counties.* Lancaster, PA: G. Hills, 1845.

Sayenga, Donald. "The Early Years of America's Wire Rope Industry 1818-1848." *Canal History and Technology Proceedings.* Vol. X, 1991.

———. "The Untryed Business: An Appreciation of White and Hazard." *Canal History and Technology Proceedings.* Vol. II, 1983.

Sears, John F. *Sacred Places: American Tourist Attractions in the Nineteenth Century.* New York: Oxford University Press, 1989; repr., University of Massachusetts Press, 1999.

Taylor, Frank H. *Autumn Leaves Upon the Lehigh.* James W. Nagle, Philadelphia, c. 1875.

Taylor, Michael. *Jane's Encyclopedia of Aviation.* Studio Editions.

Thomas Iron Company. *The Thomas Iron Company, 1854–1904.* Self published. 1904.

Vantuono, William C. "Conrail at 40: An Experiment that Worked." *Railway Age*, 5 April 2016.

Virtue, George O. "The Anthracite Mine Laborers." *Bulletin of the Department of Labor* Vol. 2, No. 13, Nov 1897.

Warren, Kenneth. *Industrial Genius: The Working Life of Charles Michael Schwab.* University of Pittsburgh Press, 2007.

Whelan, Frank. " 'I Can Stand It': Charles M. Schwab and the Bethlehem Steel strike of 1910." *Canal History and Technology Proceedings* Vol. XIII, 1994.

———. "Wire Mill Once Thrived." *The Morning Call*, 13 April 2005.

White, Josiah. *Josiah White's History, given by himself.* Lehigh Coal and Navigation Company, n.d.

Wolensky, Kenneth C., et al. *The "Great Strike": Perspectives on the 1902 Anthracite Coal Strike.* CHTP, 2002.

Wolensky, Robert P., and Joseph M. Keating. *Tragedy at Avondale.* Easton, PA: CHTP, 2008.

Wolensky, Robert P., and William A. Hastie Sr. *Anthracite Labor Wars.* Easton, PA: CHTP, 2013.

Yates, W. Ross. *Asa Packer: A Perspective.* Bethlehem, PA: Asa Packer Society, Lehigh University, 1983.

———, et al. *Bethlehem of Pennsylvania: The Golden Years, 1841–1920.* Bethlehem, PA: The Bethlehem Book Committee, 1976.

Yoder, C.P. "Bill." *Delaware Canal Journal.* Bethlehem, PA: Canal Press Inc., 1972.

Young, Joseph S. *A Brief Outline of the History of Cement.* Allentown, PA: Lehigh Portland Cement, 1942.

Zimmerman, Albright. *Pennsylvania's Delaware Division Canal: Sixty Miles of Euphoria and Frustration.* Easton, PA: CHTP, 2002.

Online sources:

http://himedo.net/TheHopkinThomasProject/TheHopThomasProject.html

http://worldpopulationreview.com/states/pennsylvania-population/

https://archive.org/stream/historyofmanufacoowill/historyofmanufacoowill_djvu.txt

https://ehistory.osu.edu/exhibitions/gildedage/1902AnthraciteStrike/content/Baer

https://www.eia.gov/coal/annual/pdf/table18.pdf

https://www.macktrucks.com/trucks/anthem/

https://www.nps.gov/parkhistory/online_books/dewa/slateford_farm/chap4.htm

www.britishpathe.com/video/another-transatlantic-plane-crash/query/01071000

www.conrail.com/history

www.globalsecurity.org/military/facility/bristol.htm

www.lhforge.com

www.livingplaces.com/PA/Bucks_County/Bristol_Borough/Harriman.html

www.nscorp.com/content/dam/nscorp/get-to-know-ns/investor-relations/annual-reports/annual-report-2017.pdf

www.phmc.state.pa.us/portal/communities/pa-heritage/files/disaster-murder-mines.pdf

www.railwayage.com/freight/class-i/conrail-at-40-the-experiment-still-works/

Index

A

Adelaide Silk Mill, 113, 164, 211
airplane production, 167–70
Allentown Iron Co., 49, 52, 88
American Cement Co., 126
American Steel & Wire Co., 163
Andrewsville, 154
anthracite, 2, 3, 4, 7
    industrial uses for, 7, 8, 37, 52, 57
    production of, 7, 33, 60, 125
    tons mined, 32, 33
    transportation of, 5, 7, 13–14, 20, 21, 29, 30, 31–32, 48, 52
anthracite iron, 169
anthracite, Wyoming, 85
anthracite production levels, 125
anthracite region, map of, 149
Anthracite Strike Commission, 150
anthracite strikes, 148
arks, coal, 7
armaments, 122, 138, 140, 143, 159, 160, 161, 162; testing of, 120
armor finishing, 121
armor plate, 106–07, 117–19, 121
Ashley Planes, 31
Atlas Cement Co., 98, 110, 127
Avondale disaster, 62

B

Baer, George, 150
Baldwin Locomotive Works, 31
Balliett, Stephen, 66, 67, 68
barbed wire, 163
battleships, 106, 162
bear trap dams, 10, 11, 19, 28
Beaver Meadow Co., 27, 32, 33
Beaver Meadow RR, 29, 30, 31
benefactors, 128–130
Bessemer steel, 77, 94, 95, 130
Bethlehem Iron Co., 49, 52, 72, 75, 76, 79, 85, 93, 95, 105, 117
    heavy forging at, 106

Bethlehem Steel Co., 138, 140–41, 159, 172, 212
Biddle, Nicholas, 36
Biery's Port, 40
bituminous coal, 3, 5, 7
blacksmiths, 2, 3, 6, 7
Blakslee, James, 59
blast furnaces, 45, 76
blasting powder, 60, 61
blowing-in, 43, 44
bonus system, 155
Bowman Brothers, 67
Boys Industrial Assn., 128, 129
Bradley Pulverizer Co., 108–09
breaker boys, 128, 153–54
breakers, coal, 61, 62, 153, 208
breweries, 137, 146, 184
Bristol, 24, 101–02, 143–44
    aircraft made at, 167–71
    historic district at, 168
    shipbuilding at, 167
Bryden Horse Shoe Works, 124
Buck Mountain Coal Co., 32, 33

C

canal boat capacity, 15, 81
canal construction, 15
canal engineers, 54
canal to railroad links, 33
canal workers, 15, 26
canals, 13, 24, 48, 86, 87, 83, 196–97
Carbon Iron Co., 67, 88
Carnegie, Andrew, 140
Carnegie Phipps, 119, 121
cartoons. Immigrants, 123; anthracite strike, 150; steel strike, 156
Catasauqua, 41, 124 206. See also Crane Iron
Catasauqua & Fogelsville RR, 64, 65
cement industry, 66, 96–97, 98, 108–12, 126–28, 179–82

cement company logos, 210
Central RR of NJ, 81, 96, 125
Chapman, Isaac, 7, 15
charcoal, 1, 3, 57
child labor, 150, 151, 153
chutes, 86
Cist, Charles, 5, 7
civil engineering, 34
coal, transporting, 3, 96, 125.
Coal and Iron Police, 150
coal regions, 32
coalfields, map of, 12
coal from other states, 33
coal preparation, 61
coalmining, methods of, 60
coke as blast furnace fuel, 87
company housing, 69, 139, 167; at Glendon, 47
Conrail locomotives, 211
consolidation of coalfields, 96
consumer spending, 1920s, 169
Cooper & Hewitt, 48, 89
Coplay, 68, 90, 97, 109–10, 112, 126
Coryell, Lewis, 25
Coxe, Eckley and Sophia, 129
Craig's Hotel, 131
Crane, George, 36, 37
Crane Iron Co., 52, 124, 172, 206. See also Lehigh Crane Co.
Crane Iron RR, 31
crusher, coal, 61
crusher, Gates, 99
crusher, Griffin, 108
cultural changes, 50
Curtiss-Wright, 170

D

dam construction, 17, 18
Darby, Abraham, 35
Davenport, Russell, 119, 138
Davies & Thomas Foundry, 124

225

Delaware, Lackawanna & Western RR, 81
Delaware Canal, 12, 17, 18, 22, 23-28, 48, 64, 81, 89, 97, 101-03, 146, 197
depression of 1870s, 90, 91
"Divine Right" letter, 150
Dixie Cup, 170-72, 212
dog house, 28
Douglas, Edwin, 27, 28, 31, 54-56
Dragon Company, 127
Drifton, 129
DuPont Co., 60, 61
Durham Furnace, 47-48, 89

E

East Pennsylvania RR, 65
Eastern Middle coal field, 33
Easton, 18, 24, 46, 55, 58, 59, 74, 104, 114, 170-71, 180, 183
Eckert, John W., 97, 98
economic downturns, 27. *See also* Panics
education, 128-29, 145, 151, 153
Egypt, 126
Ellet, Charles, 54
Emergency Fleet Corp., 166
employment figures, wartime, 163
energy crisis, 5
engineers, mechanical, 31
engineers, canal, 16, 21, 25, 27, 82
engineers, railroad, 29, 73
Evans, Oliver, 5
export of anthracite, 11

F

Fell, Jesse, 4
Ferris wheel axle, 120, 121
Firmstone, William, 44, 46, 47
Flag Staff Mountain, 116
floods, 24, 26, 28, 31, 44, 76, 78-80, 207
Fogelsville, cement in, 128
forest resources, 15
forging, heavy, 105-06, 121, 140-41, 160
forgings, large, 120
Franklin Institute award, 36
Friedensville zinc mines, 99-101
Fritz, John, 75-77, 92, 93, 106-19
fuel, furnace, 1, 2, 87

furnace described, 43, 206
furnace dimensions, 40, 42
furnaces, prototype, 37
furnaces built along railroads, 87

G

gates, lock, 27-28
geological survey, 36
Gilbert, Howard, 71
Ginder, Philip, 3
Glendon borough, 47
Glendon Iron, 44, 46-47, 52
Gore, Obediah & Daniel, 2
Gowen, Franklin, 125, 126
Grace, Eugene, 143, 172
Grant, President, 101
gravity railroads, 18, 19, 20, 31
  as tourist attraction, 114-16
grenades, 164
Grey, Henry, 142
Grey Mill, 142, 143
Grider, Rufus, 79, 94, 131
Grundy, Joseph, 145
Grundy mills, 103, 144-46

H

hammer, drop, 107
handbill, Lehigh Coal, 6, 34
hardening armor plate, 121
Harriman shipyard, 166-68, 169
Harriman, W. Averill, 166
Harvey, Hayward Augustus, 121
Hauto, George F., 10, 11
Hazard, Erskine, 7, 8, 9, 11, 23, 29, 33, 36-38, 53
Hazard, Fisher, 55, 56
Hazard Manufacturing Co., 56
Hazleton, 30, 53
Hazleton Coal Co., 32, 33
Hazleton Railroad, 31
H-beams, 142-43, 172
heavy forging, 106, 140-41
Hellertown, ore mines at, 50
hematite ore, brown, 47, 51
Hokendauqua, 41, 70
Hollenback, Matthias, 5
Holley, Alexander, 77
hospitals. Drifton, 129; Palmerton, 133, 135; South Bethlehem, 130
hot blast, 36-38

I

ice cream Dixie, 171, 212
Ihrie, Peter, 23
immigrants, 49-50, 122-25, 139
immigration, 36, 63, 122
inclined planes, 31, 54, 55, 86
industrial production, 169
industrial revolution, 1, 35-36, 44
industrialists, benevolence of, 30, 128-30
industrialization in U.S., 58
Ingham, Samuel, 23, 29
innovation, 7
iron, hot-blast, 36-38
iron, mass produced, 35, 36, 57
iron bands, 54-55
iron companies, 45, 67, 87, 172
iron furnaces, Lehigh Valley, 85, 87-91
iron ore mining, 50, 52, 64-65
iron experiments, 37
iron products, imported, 1, 36
iron-wire rope, 54
ironmasters, 75
Ironton mines, 66
Ironton Railroad, 66

J

Johnston, Archibald, 143

K

Keystone Aircraft Co., 167-70
Keystone Cement Co., 112
Keystone Furnace, 70
kiln, rotary, 110; vertical, 112

L

Lackawanna & Western RR, 126
Lackawanna Iron & Coal Co., 73
Langhorne Carpet Company, 201
Lawrence Portland Cement Co., 127
LC&N, 13, 15, 27, 29, 31, 39-40, 115; and flood of 1862, 78-80
LC&N and industrial development, 39-40, 46, 56
Leedom Carpet Co., 143-44
Leedom's coal yard, 102
legislative act to improve river, 10
Lehigh & Susquehann RR, 31-32, 58, 81, 96, 125

Lehigh Canal. *See* Lehigh Navigation
Lehigh Cement Co., 127
Lehigh coal, 125
Lehigh Coal & Navigation Co. *See* LC&N
Lehigh Coal Co., 10
Lehigh Coal Mine Co., 3-5, 7
Lehigh Crane Iron Co., 37-44, 64
Lehigh Gap, 130-131
Lehigh Heavy Forge Corp., 141
Lehigh Navigation, 12-15, 17-18, 24-28, 31, 38, 45, 46, 52, 55, 64, 71, 78, 84, 97, 125, 183, 196-97; upper section of, 27-32, 78, 81
Lehigh Navigation Company, 10
Lehigh Valley, iron production, 46, 86
Lehigh Valley Iron Co., 67-68, 90
Lehigh Valley RR, 49, 58, 73-74, 76, 81, 83, 96, 125
Lehigh Zinc, 134
Lehigh Zinc and Iron Co., 130
lift locks, 18
limonite ore, 57, 51
Livingstone Mills, 102
lock gates, 27, 28
Lock Ridge Iron Co., 70
Lower Macungie mining, 65
Lucy Furnace, 70
Lumberville, 26
Luzerne County, 60
Luzerne County philanthropists, 128, 130

M

machine shops, 160, 161
machinery, industrial, 57-58
Mack Trucks, 136-37, 164-65, 183, 193
magnetite ore, 47, 52
Mansion House Hotel, 115
manufacturing, changes in, 169
map of coal fields, xvi, 12, 22, 149
Mauch Chunk, 10, 18, 19, 31, 55, 84, 86, 205; as tourist attraction, 115-116
McLeod, Archibald, 126
Merchant Shipbuilding Co., 167
mergers, railroad, 126
mills, Bristol, 102-03
mine ownership, 52
mine safety, 62

Miner, Charles, 7
mining, anthracite, 85, 96
mining, iron ore, 50, 52
Mining and Mechanical Inst., 129-30
Mitchell, John, 150
Moore, Hugh, 170, 197
Morgan, J.P., 150
Morris Canal, 22
Mother Jones, 152
mules on gravity railroad, 20
Mumford brothers, 20
muskrats, 17

N

nationalities of immigrants, 122-25
Navarro, José de, 98, 109-12, 126
Navy, U.S., 106, 108
New Hope, 25
New Jersey Zinc Co., 130-32, 134
Norris locomotive, 33
North Pennsylvania RR, 65
Northampton borough, 97, 98, 127

O

Oliver Powder Mills, 61
open-hearth steel, 105, 107, 108
ore, iron, 52, 64-65
Ormrod, George, 127

P

Packer, Asa, 58-59, 73, 81, 130
Palmer, Bradley Webster, 129
Palmer, Ellen, 128-29
Palmer, Samuel, 131
Palmerton, 132-135
Panama Canal cement, 127
Panic of 1837, 27
Panic of 1873, 87, 91, 115
Panic of 1892, 89
Panther Valley, 115, 125, 186
paper cups, 170, 171
Pardee, Ario, 29-30
Parryville, 31, 39, 67
patch towns, 63, 154
patents for hot blast, 37, 38
Paterson, NJ, 113
Peirce, Joshua, 102
Penn Haven, 29
Pennsylvania, iron production in, 45-46

Pennsylvania and Lehigh Zinc Co., 71-72
Pennsylvania German dominance, 49, 50
Penobscot Mountain route to Wilkes-Barre, 81
Philadelphia & Reading RR, 58, 125-26, 150
philanthropy, 41, 59, 82, 92, 128-30, 131-35, 14, 197
Phoenix Silk Co., 113
pig iron, 36
Pittston, 62
planes, inclined, 31
plows, iron, 7
Plymouth Township, 4
Poco Anthracite Furnace, 67
President pump, 99
press, forging, 119, 169
prosperity, 49
puddling, 15, 17
pulverizer. *See* crusher
pump house, zinc mine, 100

Q

Quakertown & Easton Railroad, 89

R

railroad bridges, 74, 208
railroad car capacity, 81
railroad map, 125, 184, 211
railroad mergers, 126
railroads, 29, 33, 52, 85, 87, 96 125; coal carried by, 33, 48, 58, 73; cement carried by, 66; iron furnaces built along, 87
rails, iron, 36, 73, 77
rails, steel, 77
Read & Lovatt mill, 114, 133
Redington proving grounds, 160
research labs, zinc, 134
river transportation of coal, 3, 4, 7, 10-11
Roberts, Solomon, 36
Robinson, John, 7
Rodenbaugh & Stewart Co., 55
Roebling, John A., 54, 55
roller coaster, 20
Roosevelt, Theodore, and anthracite strike, 150
ropes, on inclined planes, 54, 55

S

Saucon Iron Co., 70, 91
Saucon Valley ores, 65, 66
Saucon Valley zinc ores, 71, 130
Saucona Iron Co., 75
Saylor, David, 97
Saylor Cement Co., 108, 110
Sayre, George, 55
Sayre, Robert H., 55, 73, 74, 77, 81; biography of, 82; family photo of, 83; retirement of, 138
Schoefer kilns, 126, 127
Schuylkill coal, 58
Schwab, Charles, 138, 140-43, 155, 157-59, 162; and Emergency Fleet Corp., 165, 166
shell production, WWI, 160-61
shipbuilding, WWI, 165
shipments of coal, 7, 8
ships, navy, 105
shipyards, U.S. Shipbuilding Co., 140-41
Siegfried, 127
silk mills, 104, 113-14, 176-79, 191; employment at, 149
silk ribbon for Victory Medal, 164
Simon mills, 104
smelting iron, 39
Smith, Abijah & John, 4, 5
Smith, Joseph, 6, 7
Solomon Gap tunnel, 31
South Bethlehem, 99, 138, 139
speakeasy, 139
speculation in ironmaking, 87
spiegeleisen furnaces, 134
St. Luke's hospital, 130
state police, 157
steam power, 2, 5, 23, 29, 31, 32, 33, 43, 45, 48, 54, 57, 58, 60, 67, 89, 99, 107, 108, 117-19, 125, 184.
steel plant described, 95
steel production figures, 95
steelmaking, start of, 93-95
Stoddartsville, 28
street paving, 136, 179-81
Strickland, William, 25
strikes, anthracite, 148-51; steel, 155-57
structural beams, 142, 172

subsidence, 53
Sugar Loaf Coal Co., 32, 53
Summit Hill, 3, 4, 10, 19
symbiosis, 87
Susquehanna River navigation, 32
Switchback Railroad, 20; as tourist attraction, 114-16, 209. *See also* gravity railroads
Switzerland of America, 115

T

Tamaqua, 29
Tannery, 28
Taylor, George, 40, 48
testing armaments, 120
textile industry, 1, 144
Thomas, David, 31, 36, 38-41; description of Catasauqua by, 50
Thomas, Hopkin, 29, 31
Thomas, Samuel, 40, 69
Thomas Iron Company, 41, 44, 50, 52, 64, 66, 70, 91, 172; directors of, 68; production levels of, 70
three-high mill, 75, 76
Tiffany, Louis Comfort, 94, 102
tons of coal mined, 4, 13, 32-34, 52, 57, 147, 185, 192, 202
tourism, 114-16; industrial, 20
towns, iron-company, 48
transporting coal, 5, 7, 13, 14, 20, 21, 29
Traylor, Samuel, 162, 163
Traylor Engineering, 162, 163, 166
Trexler, Harry, 127
Trojan Powder Co., 164

U

Ueberroth, Jacob, 71
Ueberroth mine, 99, 101
unions, opposition to, 30, 150-51, 155-57
United Mine Workers, 148, 150
United States Shipbuilding Corp., 140-42, 162
Upper Grand Section, 27, 28, 31, 81; and flood of 1862, 78, 80. *See also* Lehigh Navigation
U.S. Navy, 119, 120, 121

V

Vickers company, 138

W

wages, steel company, 155
Wapwallopen mills, 60-61
War of 1812, 5, 7
war production, 159-64
war workers, women, 160, 161
water in mines, 99, 101
waterpower, 1, 2, 61
Weatherly, 29, 31, 114
weigh lock, 18, 25, 37, 82
Weiss, Jacob, 3, 5
Weissport, 79
Wetherill, Samuel, 71, 130
Wharton, Joseph, 72, 138
Whitaker, Joseph and Co., 48
White, Canvass, 15, 16, 29, 96
White, Josiah, 7-9, 11, 23, 27, 29, 33, 36, 53
White Haven, 28, 29; tunnel, 31
Wilkes-Barre, 2, 7, 56, 146, 147
wire mill, 56
wire rope, 31, 53-54, 55-56
wire works, 7, 8, 163
Wolle, Augustus, 75
women workers, 114, 160, 161
wood, 3, 35
workers, Bethlehem Steel, 155, 161
working conditions, 128
World War I, 159, 161-67
Wright, Benjamin, 16, 21
Wurts brothers, 21
Wyoming Valley coal, 5, 21, 32, 52, 60, 96, 125

Y

Yniscedwyn Iron Works, 37
Young, Edward, 127

Z

Zinc, New Jersey, 130-31, 134
zinc mining, 71, 99-100
zinc, South Bethlehem, 71-72, 130